Table of Contents

About HeartMath Institute iii

Introduction 1

Chapter 1: Heart-brain Communication 3

Chapter 2: Resilience, Stress and Emotions 8

Chapter 3: Heart Rate Variability: An Indicator of Self-Regulatory Capacity,
Autonomic Function and Health 13

Chapter 4: Coherence 24

Chapter 5: Establishing a New Baseline 29

Chapter 6: Energetic Communication 36

Chapter 7: Intuition Research: Coherence and the Surprising Role of the Heart 45

Chapter 8: Health Outcome Studies 53

Chapter 9: Outcome Studies in Education 66

Chapter 10: Social Coherence: Outcome Studies in Organizations 81

Chapter 11: Global Coherence Research: Human-Earth Interconnectivity 89

Bibliography 100

© Copyright 2015 HeartMath Institute

About HeartMath Institute

HeartMath Institute (HMI) is an innovative nonprofit research and education organization that provides simple, user-friendly mental and emotion self-regulation tools and techniques that people of all ages and cultures can use in the moment to relieve stress and break through to greater levels of personal balance, stability, creativity, intuitive insight and fulfillment.

HMI research has formed the foundation for training programs conducted around the world in many different types of populations, including major corporations, government and social-service agencies, all branches of the Armed Forces, schools and universities, hospitals and a wide range of health-care professionals. The tools and technologies developed at HMI offer hope for new, effective solutions to the many daunting problems that society currently faces, beginning with restoring balance and maximizing the potential within each of us.

HeartMath Institute's Mission (HMI)

The mission of HeartMath Institute is to help people bring their physical, mental and emotional systems into balanced alignment with their heart's intuitive guidance. This unfolds the path for becoming heart-empowered individuals who choose the way of love, which they demonstrate through compassionate care for the well-being of themselves, others and Planet Earth.

© Copyright 2015 HeartMath Institute

Introduction

New research shows the human heart is much more than an efficient pump that sustains life. Our research suggests the heart also is an access point to a source of wisdom and intelligence that we can call upon to live our lives with more balance, greater creativity and enhanced intuitive capacities. All of these are important for increasing personal effectiveness, improving health and relationships and achieving greater fulfillment.

This overview will explore intriguing aspects of the science of the heart, much of which is still relatively not well known outside the fields of psychophysiology and neurocardiology. We will highlight research that bridges the science of the heart and the highly practical, research-based skill set known as the HeartMath System.

The heart has been considered the source of emotion, courage and wisdom for centuries. For more than 25 years, the HeartMath Institute Research Center has explored the physiological mechanisms by which the heart and brain communicate and how the activity of the heart influences our perceptions, emotions, intuition and health. Early on in our research we asked, among other questions, why people experience the feeling or sensation of love and other regenerative emotions as well as heartache in the physical area of the heart. In the early 1990s, we were among the first to conduct research that not only looked at how stressful emotions affect the activity in the autonomic nervous system (ANS) and the hormonal and immune systems, but also at the effects of emotions such as appreciation, compassion and care. Over the years, we have conducted many studies that have utilized many different physiological measures such as EEG (brain waves), SCL (skin conductance), ECG (heart), BP (blood pressure) and hormone levels, etc. Consistently, however, it was heart rate variability, or heart rhythms that stood out as the most dynamic and reflective indicator of one's emotional states and, therefore, current stress and cognitive processes. It became clear that stressful or depleting emotions such as frustration and overwhelm lead to increased disorder in the higher-level brain centers and autonomic nervous system and which are reflected in the heart rhythms and adversely affects the functioning of virtually all bodily systems. This eventually led to a much deeper understanding of the neural and other communication pathways between the heart and brain. We also observed that the heart acted as though it had a mind of its own and could significantly influence the way we perceive and respond in our daily interactions. In essence, it appeared that the heart could affect our awareness, perceptions and intelligence. Numerous studies have since shown that heart coherence is an optimal physiological state associated with increased cognitive function, self-regulatory capacity, emotional stability and resilience.

We now have a much deeper scientific understanding of many of our original questions that explains how and why heart activity affects mental clarity, creativity, emotional balance, intuition and personal effectiveness. Our and others' research indicates the heart is far more than a simple pump. The heart is, in fact, a highly complex information-processing center with its own functional brain, commonly called the *heart brain*, that communicates with and influences the cranial brain via the nervous system, hormonal system and other pathways. These influences affect brain function and most of the body's major organs and play an important role in mental and emotional experience and the quality of our lives.

In recent years, we have conducted a number of research studies that have explored topics such as the electrophysiology of intuition and the degree to which the heart's magnetic field, which radiates outside the body, carries information that affects other people and even our pets, and links people together in surprising ways. We also launched the Global Coherence Initiative (GCI), which explores the interconnectivity of humanity with Earth's magnetic fields.

This overview discusses the main findings of our research and the fascinating and important role the heart plays in our personal coherence and the positive changes that occur in health, mental functions, perception, happiness and energy levels as people practice the HeartMath techniques. Practicing the techniques increases heart coherence and one's ability to self-regulate emotions from a more intuitive, intelligent and balanced inner reference. This also explains how coherence is reflected in our physiology and can be objectively measured.

The discussion then expands from physiological coherence to coherence in the context of families, workplaces and communities. *Science of the Heart* concludes with the perspective that being responsible for and increasing our personal coherence not only improves personal health and happiness, but also feeds into and influences a global field environment. It is postulated that as increasing numbers of people add coherent energy to the global field, it helps strengthen and stabilize mutually beneficial feedback loops between human beings and Earth's magnetic fields.

CHAPTER 1

Heart-Brain Communication

Traditionally, the study of communication pathways between the head and heart has been approached from a rather one-sided perspective, with scientists focusing primarily on the heart's responses to the brain's commands. We have learned, however, that communication between the heart and brain actually is a dynamic, ongoing, two-way dialogue, with each organ continuously influencing the other's function. Research has shown that the heart communicates to the brain in four major ways: *neurologically* (through the transmission of nerve impulses), biochemically (via hormones and neurotransmitters), biophysically (through pressure waves) and energetically (through electromagnetic field interactions). Communication along all these conduits significantly affects the brain's activity. Moreover, our research shows that messages the heart sends to the brain also can affect performance.

> **The heart communicates with the brain and body in four ways:**
> - Neurological communication (**nervous system**)
> - Biochemical communication (**hormones**)
> - Biophysical communication (**pulse wave**)
> - Energetic communication (**electromagnetic fields**)

Some of the first researchers in the field of psychophysiology to examine the interactions between the heart and brain were John and Beatrice Lacey. During 20 years of research throughout the 1960s and '70s, they observed that the heart communicates with the brain in ways that significantly affect how we perceive and react to the world.

In physiologist and researcher Walter Bradford Cannon's view, when we are aroused, the mobilizing part of the nervous system (sympathetic) energizes us for fight or flight, which is indicated by an increase in heart rate, and in more quiescent moments, the calming part of the nervous system (parasympathetic) calms us down and slows the heart rate. Cannon believed the autonomic nervous system and all of the related physiological responses moved in concert with the brain's response to any given stimulus or challenge. Presumably, all of our inner systems are activated together when we are aroused and calm down together when we are at rest and the brain is in control of the entire process. Cannon also introduced the concept of homeostasis. Since then, the study of physiology has been based on the principle that all cells, tissues and organs strive to maintain a static or constant steady-state condition. However, with the introduction of signal-processing technologies that can acquire continuous data over time from physiological processes such as heart rate (HR), blood pressure (BP) and nerve activity, it has become abundantly apparent that biological processes vary in complex and nonlinear ways, even during so-called steady-state conditions. These observations have led to the understanding that healthy, optimal function is a result of continuous, dynamic, bidirectional interactions among multiple neural, hormonal and mechanical control systems at both local and central levels. In concert, these dynamic and interconnected physiological and psychological regulatory systems are never truly at rest and are certainly never static.

For example, we now know that the normal resting *rhythm* of the heart is highly variable rather than monotonously regular, which was the widespread notion for many years. This will be discussed further in the section on heart rate variability (HRV).

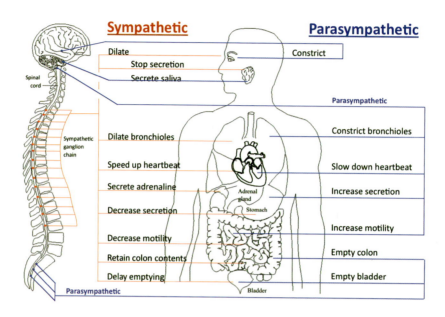

Figure 1.1 Innervation of the major organs by the autonomic nervous system (ANS). Parasympathetic fibers are primarily in the vagus nerves, but some that regulate subdiaphragmatic organs travel through the spinal cord. The sympathetic fibers also travel through the spinal cord. A number of health problems can arise in part because of improper function of the ANS. Emotions can affect activity in both branches of the ANS. For example, anger causes increased sympathetic activity while many relaxation techniques increase parasympathetic activity.

The Laceys noticed that the model proposed by Cannon only partially matched actual physiological behavior. As their research evolved, they found that the heart in particular seemed to have its own logic that frequently diverged from the direction of autonomic nervous system activity. The heart was behaving as though it had a mind of its own. Furthermore, the heart appeared to be sending meaningful messages to the brain that the brain not only understood, but also obeyed. Even more intriguing was that it looked as though these messages could affect a person's perceptions, behavior and performance. The Laceys identified a neural pathway and mechanism whereby input from the heart to the brain could inhibit or facilitate the brain's electrical activity. Then in 1974, French researchers stimulated the vagus nerve (which carries many of the signals from the heart to the brain) in cats and found that the brain's electrical response was reduced to about half its normal rate.[1] This suggested that the heart and nervous system were not simply following the brain's directions, as Cannon had thought. Rather, the autonomic nervous system and the communication between the heart and brain were much more complex, and the heart seemed to have its own type of logic and acted independently of the signals sent from the brain.

While the Laceys research focused on activity occurring within a single cardiac cycle, they also were able to confirm that cardiovascular activity influences perception and cognitive performance, but there were still some inconsistencies in the results. These inconsistencies were resolved in Germany by Velden and Wölk, who later demonstrated that cognitive performance fluctuated at a rhythm around 10 hertz throughout the cardiac cycle. They showed that the modulation of cortical function resulted from ascending cardiovascular inputs on neurons in the thalamus, which globally synchronizes cortical activity.[2, 3] An important aspect of their work was the finding that it is the pattern and stability of the heart's rhythm of the afferent (ascending) inputs, rather than the number of neural bursts within the cardiac cycle, that are important in modulating thalamic activity, which in turn has global effects on brain function. There has since been a growing body of research indicating that afferent information processed by the intrinsic cardiac nervous system (heart-brain) can influence activity in the frontocortical areas[4-6] and motor cortex,[7] affecting psychological factors such as attention level, motivation,[8] perceptual sensitivity[9] and emotional processing.[10]

Neurocardiology: The Brain On the Heart

While the Laceys were conducting their research in psychophysiology, a small group of cardiologists joined forces with a group of neurophysiologists and neuroanatomists to explore areas of mutual interest. This represented the beginning of the new discipline now called *neurocardiology*. One of their early findings is that the heart has a complex neural network that is sufficiently extensive to be characterized as a brain on the heart (Figure 1.2).[11, 12] The *heart-brain*, as it is commonly called, or intrinsic cardiac nervous system, is an intricate network of complex ganglia, neurotransmitters, proteins and support cells, the same as those of the brain in the head. The heart-brain's neural circuitry enables it to act independently of the cranial brain to learn, remember, make decisions and even feel and sense. Descending activity from the brain in the head via the sympathetic and parasympathetic branches of the ANS is integrated into the heart's intrinsic nervous system along with signals arising from sensory neurons in the heart that detect pressure, heart rate, heart rhythm and hormones.

The anatomy and functions of the intrinsic cardiac nervous system and its connections with the brain have been explored extensively by neurocardiologists.[13, 14] In terms of heart-brain communication, it is generally well-known that the efferent (descending) pathways in the autonomic nervous system are involved in the regulation of the heart. However, it is less appreciated that the majority of fibers in the vagus nerves are afferent (ascending) in nature. Furthermore, more of these ascending neural pathways are related to the heart (and cardiovascular system) than to any other organ.[15] This means the heart sends more information to the brain than the brain sends to the heart. More recent research shows that the neural interactions between the heart and brain are more complex than previously thought. In addition, the intrinsic cardiac nervous system has both short-term and long-term memory functions and can operate independently of central neuronal command.

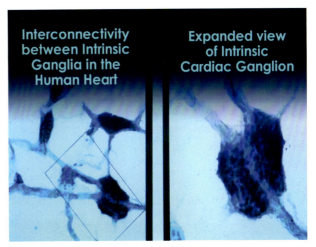

Figure 1.2. Microscopic image of interconnected intrinsic cardiac ganglia in the human heart. The thin, light-blue structures are multiple axons that connect the ganglia. Courtesy of Dr. J. Andrew Armour.

Once information has been processed by the heart's intrinsic nervous system, the appropriate signals are sent to the heart's sinoatrial node and to other tissues in the heart. Thus, under normal physiological conditions, the heart's intrinsic nervous system plays an important role in much of the routine control of cardiac function, independent of the central nervous system. The heart's intrinsic nervous system is vital for the maintenance of cardiovascular stability and efficiency and without it, the heart cannot function properly. The neural output, or messages from the intrinsic cardiac nervous system travels to the brain via ascending pathways in the both the spinal column and vagus nerves, where it travels to the medulla, hypothalamus, thalamus and amygdala and then to the cerebral cortex.[5, 16, 17] The nervous-system pathways between the heart and brain are shown in Figure 1.3 and the primary afferent pathways in the brain are shown in Figure 1.4.

Had the existence of the intrinsic cardiac nervous system and the complexity of the neural communication between the heart and brain been known while the Laceys were conducting their paradigm-shifting research, their theories and data likely would have been accepted far sooner. Their insight, rigorous experimentation and courage to follow where the data led them, even though it did not fit the well-entrenched beliefs of the scientific community of their day, were pivotal in the understanding of the heart-brain connection. Their research played an important role in elucidating the basic physiological and psychological processes that connect the heart and brain and the mind and body. In 1977, Dr. Francis Waldropin, director of the National Institute of Mental Health, stated in a review article of the Laceys' work, "Their intricate and careful procedures, combined with their daring theories, have produced work that has stirred controversy as well as promise. In the long run, their research may tell us much about what makes each of us a whole person and may suggest techniques that can restore a distressed person to health."

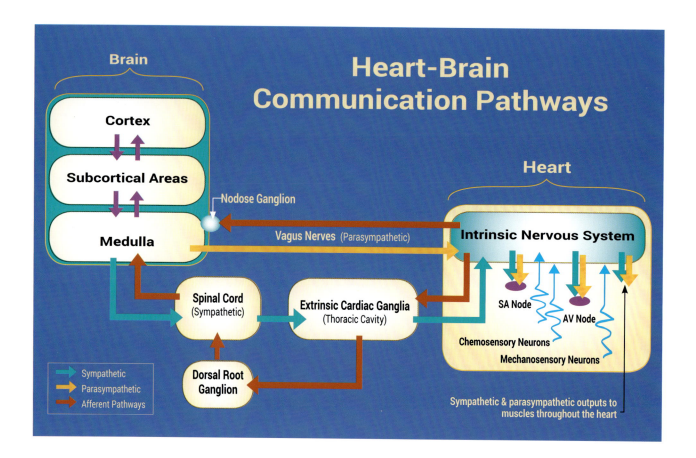

Figure 1.3. The neural communication pathways interacting between the heart and brain are responsible for the generation of HRV. The intrinsic cardiac nervous system integrates information from the extrinsic nervous system and the sensory neurites within the heart. The extrinsic cardiac ganglia located in the thoracic cavity have connections to the lungs and esophagus and are indirectly connected via the spinal cord to many other organs, including the skin and arteries. The vagus nerve (parasympathetic) primarily consists of afferent (flowing to the brain) fibers that connect to the medulla. The sympathetic afferent nerves first connect to the extrinsic cardiac ganglia (also a processing center), then to the dorsal root ganglion and the spinal cord. Once afferent signals reach the medulla, they travel to the subcortical areas (thalamus, amygdala, etc.) and then the higher cortical areas.

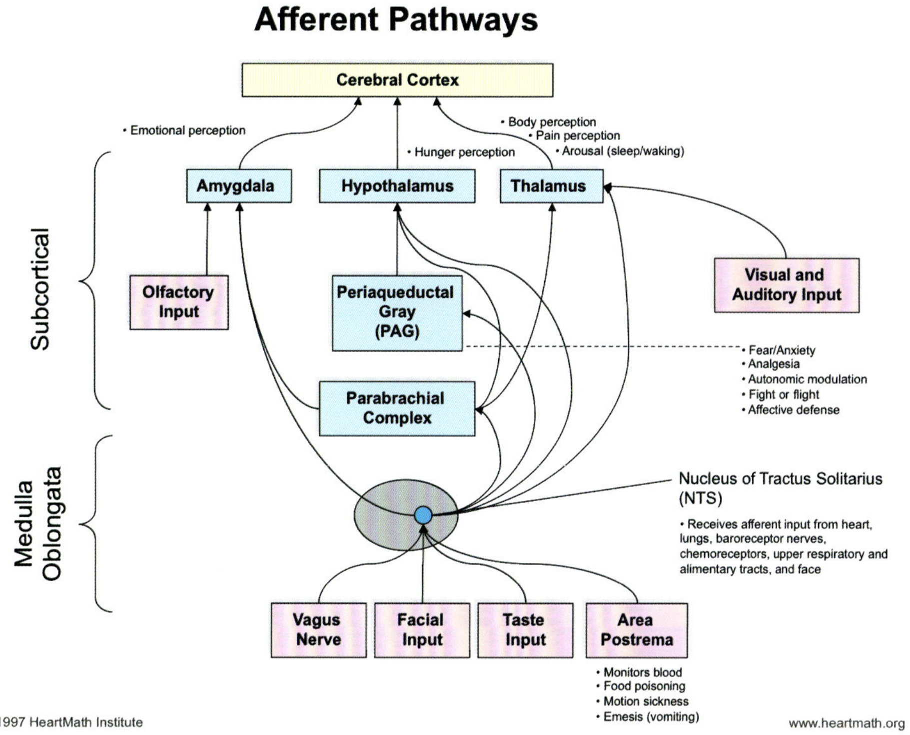

Afferent Pathways

Figure 1.4. Diagram of the currently known afferent pathways by which information from the heart and cardiovascular system modulates brain activity. Note the direct connections from the NTS to the amygdala, hypothalamus and thalamus. Although not shown, there also is evidence emerging that there is a pathway from the dorsal vagal complex that travels directly to the frontal cortex.

The Heart as a Hormonal Gland

In addition to its extensive neurological interactions, the heart also communicates with the brain and body biochemically by way of the hormones it produces. Although not typically thought of as an endocrine gland, the heart actually manufactures and secretes a number of hormones and neurotransmitters that have a wide-ranging impact on the body as a whole.

The heart was reclassified as part of the hormonal system in 1983, when a new hormone produced and secreted by the atria of the heart was discovered. This hormone has been called by several different names – atrial natriuretic factor (ANF), atrial natriuretic peptide (ANP) and atrial peptide. Nicknamed the balance hormone, it plays an important role in fluid and electrolyte balance and helps regulate the blood vessels, kidneys, adrenal glands and many regulatory centers in the brain.[18] Increased atrial peptide inhibits the release of stress hormones,[19] reduces sympathetic outflow[20] and appears to interact with the immune system.[21] Even more intriguing, experiments suggest atrial peptide can influence motivation and behavior.[22]

It was later discovered the heart contains cells that synthesize and release catecholamines (norepinephrine, epinephrine and dopamine), which are neurotransmitters once thought to be produced only by neurons in the brain and ganglia.[23] More recently, it was discovered the heart also manufactures and secretes oxytocin, which can act as a neurotransmitter and commonly is referred to as the love or social-bonding hormone. Beyond its well-known functions in childbirth and lactation, oxytocin also has been shown to be involved in cognition, tolerance, trust and friendship and the establishment of enduring pair-bonds. Remarkably, concentrations of oxytocin produced in the heart are in the same range as those produced in the brain.[24]

CHAPTER 2
Resilience, Stress and Emotions

As far back as the middle of the last century, it was recognized that the heart, overtaxed by constant emotional influences or excessive physical effort and thus deprived of its appropriate rest, suffers disorders of function and becomes vulnerable to disease.[25]

An early editorial on the relationships between stress and the heart accepted the proposition that in about half of patients, strong emotional upsets precipitated heart failure. Unspecified negative emotional arousal, often described as stress, distress or upset, has been associated with a variety of pathological conditions, including hypertension,[26, 27] silent myocardial ischemia,[28] sudden cardiac death,[29] coronary disease,[30-32] cardiac arrhythmia,[33] sleep disorders,[34] metabolic syndrome,[35] diabetes,[36, 37] neurodegenerative diseases,[38] fatigue[39, 40] and many other disorders.[41] Stress and negative emotions have been shown to increase disease severity and worsen prognosis for individuals suffering from a number of different pathologies.[42, 43] On the other hand, positive emotions and effective emotion self-regulation skills have been shown to prolong health and significantly reduce premature mortality.[44-49] From a psychophysiological perspective, emotions are central to the experience of stress. It is the feelings of anxiety, irritation, frustration, lack of control, and hopelessness that are actually what we experience when we describe ourselves as stressed. Whether it's a minor inconvenience or a major life change, situations are experienced as stressful to the extent that they trigger emotions such as annoyance, irritation, anxiety and overwhelm.[50]

In essence, stress is emotional unease, the experience of which ranges from low-grade feelings of emotional unrest to intense inner turmoil. Stressful emotions clearly can arise in response to external challenges or events, and also from ongoing internal dialogs and attitudes. Recurring feelings of worry, anxiety, anger, judgment, resentment, impatience, overwhelm and self-doubt often consume a large part of our energy and dull our day-to-day life experiences.

Additionally, emotions, much more so than thoughts alone, activate the physiological changes comprising the stress response. Our research shows a purely mental activity such as cognitively recalling a past situation that provoked anger does not produce nearly as profound an effect on physiological processes as actually engaging the emotion associated with that memory. In other words, reexperiencing the feeling of anger provoked by the memory has a greater effect than thinking about it.[51, 52]

Resilience and Emotion Self-Regulation

Our emotions infuse life with a rich texture and transform our conscious experience into a meaningful living experience. Emotions determine what we care about and what motivates us. They connect us to others and give us the courage to do what needs to be done, to appreciate our successes, to protect and support the people we love and have compassion and kindness for those who are in need of our help. Emotions are also what allow us to experience the pain and grief of loss. Without emotions, life would lack meaning and purpose.

Emotions and resilience are closely related because emotions are the primary drivers of many key physiological processes involved in energy regulation. We define resilience as *the capacity to prepare for, recover from and adapt in the face of stress, adversity, trauma or challenge.*[53] Therefore, it follows that a key to sustaining good health, optimal function and resilience is the ability to manage one's emotions.

It has been suggested that resilience should be considered as a state rather than a trait and that a person's resilience can vary over time as demands, circumstances and level of maturity change.[54] In our resilience training programs, we suggest that the ability to build and sustain resilience is related to self-management and efficient utilization of energy resources across four domains: *physical, emotional, mental and spiritual* (Figure 2.1). Physical resilience is basically reflected in physical flexibility, endurance and strength, while emotional resilience is reflected in the ability to self-regulate, degree of emotional flexibility, positive outlook and supportive relationships. Mental resilience is reflected in the ability to sustain focus and attention, mental flexibility and the capacity for integrating multiple points of view. Spiritual resilience is typically associated with commitment to core values, intuition and tolerance of others' values and beliefs.

Figure 2.1. Domains of Resilience.

By learning self-regulation techniques that allow us to shift our physiology into a more coherent state, the increased physiological efficiency and alignment of the mental and emotional systems accumulates resilience (energy) across all four energetic domains. Having a high level of resilience is important not only for bouncing back from challenging situations, but also for preventing unnecessary stress reactions (frustration, impatience, anxiety), which often lead to further energy and time waste and deplete our physical and psychological resources.

Most people would agree it is the ability to adjust and self-regulate one's responses and behavior that is most important in building and maintaining supportive, loving relationships and effectively meeting life's demands with composure, consistency and integrity. The ability to adjust and self-regulate also is central to resilience, good health and effective decision-making.[55] It is a key for success in living life with greater kindness and compassion in all relationships. If people's capacity for intelligent, self-directed regulation is strong enough, then regardless of inclinations, past experiences or personality traits, they usually can do the adaptive or right thing in most situations.[56]

We are coming to understand health not as the absence of disease, but rather as the process by which individuals maintain their sense of coherence (i.e. sense that life is comprehensible, manageable, and meaningful) and ability to function in the face of changes in themselves and their relationships with their environment.[57]

It has been shown that our efforts to self-regulate emotions can produce broad improvements in increasing or strengthening self-regulatory capacity, similar to the process of strengthening a muscle, making us less vulnerable to depletion of our internal reserves.[56] When internal energy reserves are depleted, normal capacity to maintain self-control is weakened, which can lead to increased stress, inappropriate behaviors, lost opportunities, poor communication and damaged relationships. Despite the importance of self-directed control, many people's ability to self-regulate is far less than ideal. In fact, failures in self-regulation, especially of emotions and attitudes, arguably are central to the vast majority of personal and social problems that plague modern societies. For some, the lack of self-regulatory capacity can be attributed to immaturity or failure to acquire skills while for others it can be the result of trauma or impairment in the neural systems that underlie one's ability to self-regulate.[58] Therefore, we submit the most important skill the majority of people need to learn is how to increase their capacity to self-regulate emotions, attitudes and behaviors. Self-regulation enables people to mature and meet the challenges and stresses of everyday life with resilience so they can make more intelligent decisions by aligning with their innate higher-order wisdom and expression of care and compassion, elements we often associate with living a more conscientious life.

Our research suggests a new inner baseline reference can be established by using the HeartMath (HM) self-regulation techniques that help people replace depleting emotional undercurrents with more positive, regenerative attitudes, feelings and perceptions. This new baseline, which will be summarized in a later section, can be thought of as a type of implicit memory that organizes perception, feelings and behavior.[5, 59] The process of establishing a new baseline takes place at the physiological level, which is imperative for sustained and lasting change to occur.

> **A growing body of compelling scientific evidence is demonstrating a link between mental and emotional attitudes, physiological health and long-term well-being:**
>
> *60% to 80% of primary care doctor visits are related to stress, yet only 3% of patients receive stress management help.*[60-62]
>
> *In a study of 5,716 middle-aged people, those with the highest self-regulation abilities were over 50 times more likely to be alive and without chronic disease 15 years later than those with the lowest self-regulation scores.*[63]
>
> *Positive emotions are a reliable predictor of better health, even for those without food or shelter while negative emotions are a reliable predictor of worse health even when basic needs like food, shelter and safety are met.*[64]
>
> *A Harvard Medical School Study of 1,623 heart attack survivors found that when subjects became angry during emotional conflicts, their risk of subsequent heart attacks was more than double that of those who remained calm.*[65]
>
> *A review of 225 studies concluded that positive emotions promote and foster sociability and activity, altruism, strong bodies and immune systems, effective conflict resolution skills, success and thriving.*[66]
>
> *A study of elderly nuns found that those who expressed the most positive emotions in early adulthood lived an average of 10 years longer.*[67]
>
> *Men who complain of high anxiety are up to six times more likely than calmer men to suffer sudden cardiac death.*[68]
>
> *In a groundbreaking study of 1,200 people at high risk of poor health, those who learned to alter unhealthy mental and emotional attitudes through self-regulation training were over four times more likely to be alive 13 years later than an equal-sized control group.*[69]
>
> *A 20-year study of over 1,700 older men conducted by the Harvard School of Public Health found that worry about social conditions, health and personal finances all significantly increased the risk of coronary heart disease.*[70]
>
> *Over one-half of heart disease cases are not explained by the standard risk factors such as high cholesterol, smoking or sedentary lifestyle.*[71]
>
> *An international study of 2,829 people ages 55 to 85 found that individuals who reported the highest levels of personal mastery – feelings of control over life events – had a nearly 60% lower risk of than those who felt relatively helpless in the face of life's challenges.*[72]
>
> *According to a Mayo Clinic study of individuals with heart disease, psychological stress was the strongest predictor of future cardiac events such as cardiac death, cardiac arrest and heart attacks.*[73]
>
> *Three 10-year studies concluded that emotional stress was more predictive of death from cancer and cardiovascular disease than from smoking; people who were unable to effectively manage their stress had a 40% higher death rate than nonstressed individuals.*[74]
>
> *A study of heart attack survivors showed that patients' emotional states and relationships in the period after myocardial infarction were as important as the disease severity in determining their prognosis.*[75]
>
> *Separate studies showed that the risk of developing heart disease is significantly increased for people who impulsively vent their anger as well as for those who tend to repress angry feelings.*[76, 77]

Cognitive and Emotional System Integration

Dating back to the ancient Greeks, human thinking and feeling, intellect and emotion have been considered separate functions. These contrasting aspects of the soul, as the Greeks called them, often have been portrayed as being engaged in a constant battle for control of the human psyche. In Plato's view, emotions were like wild horses that had to be reined in by the intellect and willpower.

Research in neuroscience confirms that emotion and cognition can best be thought of as separate but interacting functions and systems that communicate via bidirectional neural connections between the neocortex, the body and emotional centers such as the amygdala and body.[78] These connections allow emotion-related input to modulate cortical activity while cognitive input from the cortex modulates emotional processing. However, the neural connections that transmit information from the emotional centers to the cognitive centers in the brain are stronger and more numerous than those that convey information from the cognitive to the emotional centers. This fundamental asymmetry accounts for the powerful influence of input from the emotional system on cognitive functions such as attention, perception and memory as well as higher-order thought processes. Conversely, the comparatively limited influence of input from the cognitive system on emotional processing helps explain why it is generally difficult to willfully modulate emotions through thought alone.

There can be differences from one individual to the next in these reciprocal connections and interactions between the cognitive and emotional systems that affect the way we perceive, experience and eventually remember our emotional experiences, and how we respond to emotionally challenging situations. Unbalanced interactions between the emotional and cognitive systems can lead to devastating effects such as those observed in mood and anxiety disorders.[78]

Although there has been a historical bias favoring the viewpoint that emotions interfere with and can be at odds with rational thinking, which of course can occur in some cases, emotions have their own type of rationality and have been shown to be critical in decision-making.[79] For example, Damasio points out, patients with damage in areas of the brain that integrate the emotional and cognitive systems can no longer effectively function in the day-to-day world, even though their mental abilities are perfectly normal. In the mid-1990s, the concept of emotional intelligence was introduced, precipitating persuasive arguments that the viewpoint of human intelligence being essentially mind intellect was far too narrow. This was because it ignored a range of human capacities that bear equal if not greater weight in determining our successes in life. Qualities such as self-awareness, motivation, altruism and compassion, but especially one's ability to self-regulate and control impulses and self-direct emotions were found to be as important or more important than a high IQ. Those qualities, more so than IQ, enable people to excel in the face of life's challenges.[80]

It is our experience that the degree of alignment between the mind and emotions can vary considerably. When they are out of sync, it can result in radical behavior changes that cause us to feel like there are two different people inside the same body. It can also result in confusion, difficulty in making decisions, anxiety and a lack of alignment with our deeper core values. Conversely, when the mind and emotions are in sync, we are more self-secure and aligned with our deeper core values and respond to stressful situations with increased resilience and inner balance.

Our research indicates that the key to the successful integration of the mind and emotions lies in increasing one's emotional self-awareness and the coherence of, or harmonious function and interaction among, the neural systems that underlie cognitive and emotional experience.[5, 58, 81]

As will be discussed in more detail in a later section, we use the terms *cardiac coherence, physiological coherence* and *heart coherence* interchangeably to describe the measurement of the order, stability and harmony in the oscillatory outputs of the body's regulatory systems during any period of time.

An important aspect of understanding how to increase self-regulatory capacity and the balance between the cognitive and emotional systems is the inclusion of the heart's ascending neuronal inputs on subcortical (emotional) and cortical (cognitive) structures which, as discussed above, can have significant influences on cognitive resources and emotions. Information is conveyed in the patterns of the heart's rhythms (HRV), that reflects current emotional states. The patterns of afferent neural input (coherence and incoherence) to the brain affect emotional experience and modulate cortical function and self-regulatory capacity. We have found that intentional activation of positive emotions plays an important role in increasing cardiac coherence and thus self-regulatory capacity.[5] These findings expand on a large body of research into the ways positive emotional states can benefit physical, mental and emotional health.[44-49]

Because emotions exert such a powerful influence on cognitive activity, intervening at the emotional level is often the most efficient way to initiate change in mental patterns and processes. Our research demonstrates that the application of emotion self-regulation techniques along with the use of facilitative technology (emWave®, Inner Balance™) can help people bring the heart, mind and emotions into greater alignment. Greater alignment is associated with improved decision-making, creativity, listening ability, reaction times and coordination and mental clarity.[81]

CHAPTER 3

Heart Rate Variability: An Indicator of Self-Regulatory Capacity, Autonomic Function and Health

The autonomic nervous system (ANS) (Figure 1.1) is the part of the nervous system that controls the body's internal functions, including heart rate, gastrointestinal tract and secretions of many glands. The ANS also controls many other vital activities such as respiration, and it interacts with immune and hormonal system functions. It is well known that mental and emotional states directly affect activity in the ANS.

The autonomic nervous system must be considered as a complex system in which both efferent (descending) and afferent (ascending) vagal (parasympathetic) neurons regulate adaptive responses. Considerable evidence suggests evolution of the ANS, specifically the vagus nerves, was central to development of emotional experience, the ability to self-regulate emotional processes and social behavior and that it underlies the social engagement system. As human beings, we are not limited to fight, flight, or freeze responses. We can self-regulate and initiate pro-social behaviors when we encounter challenges, disagreements and stressors. The healthy function of the social engagement system depends upon the proper functioning of the vagus nerves, which act as a vagal brake. This system underlies one's ability to self-regulate and calm oneself by inhibiting sympathetic outflow to targets like the heart and adrenal glands. This implies that measurements of vagal activity could serve as a marker for one's ability to self-regulate. This also suggests that the evolution and healthy function of the ANS determines the boundaries for the range of one's emotional expression, quality of communication and the ability to self-regulate emotions and behaviors.[82]

Many of HMI's research studies have examined the influence of emotions on the ANS utilizing analysis of heart rate variability/heart rhythms, which reflects heart-brain interactions and autonomic nervous system dynamics.[5, 83]

The investigation of the heart's complex rhythms, or HRV began with the emergence of modern signal processing in the 1960s and 1970s and has rapidly expanded in more recent times.[84] The irregular behavior of the heartbeat is readily apparent when heart rate is examined on a beat-to-beat basis, but is overlooked when a mean value over time is calculated. These fluctuations in heart rate result from complex, nonlinear interactions among a number of different physiological systems (Figure 3.1).

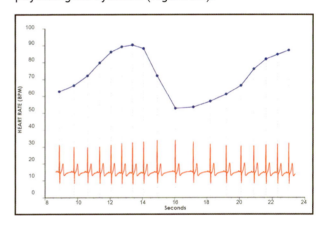

Figure 3.1. Heart rate variability is a measure of the normally occurring beat-to-beat changes in heart rate. The electrocardiogram (ECG) is shown on the bottom and the instantaneous heart rate is shown by the blue line. The time between each of the heartbeats (blue line) between 0 and approximately 13 seconds becomes progressively shorter and heart rate accelerates and then starts to decelerate around 13 seconds. This pattern of heart-rate accelerations and decelerations is the basis of the heart's rhythms.

An optimal level of HRV within an organism reflects healthy function and an inherent self-regulatory capacity, adaptability, and resilience.[5, 58, 59, 85-88] While too much instability, such as arrhythmias or nervous system chaos, is detrimental to efficient physiological functioning and energy utilization, too little variation indicates age-related system depletion, chronic stress, pathology or inadequate functioning in various levels of self-regulatory control systems.[84, 89, 90]

The importance of HRV as an index of the functional status of physiological control systems was noted as far back as 1965, when it was found that fetal distress was preceded by reductions in HRV before any changes occurred in heart rate.[91] In the 1970s, reduced HRV was shown to predict autonomic neuropathy in diabetic patients before the onset of symptoms.[92-94] Reduced HRV also was found to be a higher risk factor of death post-myocardial infarction than other known risk factors.[95] It has been shown that HRV declines with age and that age-adjusted values should be used in the context of risk prediction.[96] Age-adjusted HRV that is low has been confirmed as a strong, independent predictor of future health problems in both healthy people and in patients with known coronary artery disease and correlates with all-cause mortality.[97, 98]

Based on indirect evidence, reduced HRV may correlate with disease and mortality because it reflects reduced regulatory capacity and ability to adapt/respond to physiological challenges such as exercise. For example, in the *Chicago Health, Aging and Social Relations Study*, separate metrics for the assessment of autonomic balance and overall cardiac autonomic regulation were developed and tested in a sample of 229 participants. In this study, overall regulatory capacity was a significant predictor of overall health status, but autonomic balance was not. In addition, cardiac regulatory capacity was negatively associated with the prior incidence of myocardial infarctions. The authors suggest that cardiac regulatory capacity reflects a physiological state that is more relevant to health than the independent sympathetic or parasympathetic controls, or the autonomic balance between these controls as indexed by different measures of HRV.[99]

Heart rate variability also indicates psychological resiliency and behavioral flexibility, reflecting an individual's capacity to self-regulate and effectively adapt to changing social or environmental demands.[99, 100] A growing number of studies have specifically linked vagally mediated HRV to self-regulatory capacity,[87, 88, 101] emotional regulation,[102, 103] social interactions,[86, 104] one's sense of coherence[105] and the personality character traits of self-directedness[106] and coping styles.[107]

More recently, several studies have shown an association between higher levels of resting HRV and performance on cognitive performance tasks requiring the use of executive functions.[89] HRV coherence (described later) can be increased in order to improve cognitive function[5, 108-110] as well as a wide range of clinical outcomes, including reduced health-care costs.[59, 111-116]

Self-Regulation: Cortical Systems

Considerable evidence from clinical, physiological and anatomical research has identified cortical, subcortical and medulla oblongata structures involved in cardiac regulation. Oppenheimer and Hopkins mapped a detailed hierarchy of cardiac control structures among the cortex, amygdala and other subcortical structures, all of which can modify cardiovascular-related neurons in the lower levels of the neuraxis (Figure 3.2).[117]

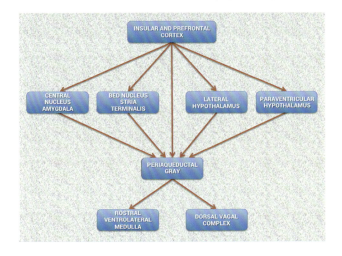

Figure 3.2. Schematic diagram showing the relationship of the principal descending neural pathways from the insular and prefrontal cortex to subcortical structures and the medulla oblongata as outlined by Oppenheimer and Hopkins.[117] The insular and prefrontal cortexes are key sites involved in modulating the heart's rhythm, particularly during emotionally charged circumstances. These structures along with other centers such as the orbitofrontal cortex and cingulate gyrus can inhibit or enhance emotional responses. The amygdala is involved with refined integration of emotional content in higher centers to produce cardiovascular responses that are appropriate for the emotional aspects of current circumstances. Imbalances between the neurons in the insula, amygdala and hypothalamus may initiate cardiac rhythm disturbances and arrhythmias. The structures in the medulla represent an interface between incoming afferent information from the heart, lungs and other bodily systems and outgoing efferent neuronal activity.[117]

They suggest that the amygdala is involved with refined integration of emotional content in higher centers to produce cardiovascular responses that are appropriate for the emotional aspects of the current circumstances. The insular cortex and other centers such as the orbitofrontal cortex and cingulate gyrus can overcome (self-regulate) emotionally entrained responses by inhibiting or enhancing them. They also point out that imbalances between the neurons in the insula, amygdala and hypothalamus may initiate cardiac rhythm disturbances and arrhythmias. The data suggests that the insular and medial prefrontal cortexes are key sites involved in modulating the heart's rhythm, particularly during emotionally charged circumstances.

Thayer and Lane also have described the same set of neural structures outlined by Oppenheimer and Hopkins, which they call the central autonomic network (CAN). The CAN is involved in cognitive, emotional and autonomic regulation, which they linked directly to HRV and cognitive performance. In their model, the CAN links the nucleus of tractus solitarius in the medulla with the insula, prefrontal cortex, amygdala and hypothalamus through a series of feedback and feedforward loops. They also propose that this network is an integrated system for internal self-regulation by which the brain controls the heart and other internal organs, neuroendocrine and behavioral responses that are critical for goal-directed behavior, adaptability and sustained health. They suggest that these dynamic connections explain why parasympathetically (vagal) mediated HRV is linked to higher-level executive functions and reflects the functional capacity of the brain structures that support working memory and emotional and physiological self-regulation. They have shown that higher levels of vagally mediated HRV are correlated with prefrontal cortical performance and the ability to inhibit unwanted memories and intrusive thoughts. The prefrontal cortex can be taken offline when individuals perceive that they are threatened, and prolonged periods of prefrontal cortical inactivity can lead to hypervigilance, defensiveness and social isolation. During these decreases in prefrontal cortical activation, heart rate (HR) increases and HRV decreases.[89]

> - *Thoughts and even subtle emotions influence the activity in the autonomic nervous system.*
> - *The ANS interacts with our digestive, cardiovascular, immune, hormonal and many other bodily systems.*
> - *Negative emotions/feelings create disorder in the brain's regulatory systems and ANS.*
> - *Feelings such as appreciation create increased order in the brain's regulatory systems and ANS, resulting in improved hormonal- and immune-system function and enhanced cognitive function.*

The nucleus of tractus in the medulla oblongata integrates afferent sensory information from proprioceptors (body position), chemoreceptors (blood chemistry) and mechanoreceptors, also called baroreceptors, (pressure or distortion) from the heart, lungs and face. The nucleus of tractus connects to the dorsal motor nucleus of the vagus nerve and the nucleus ambiguus. Neurocardiology research indicates that the descending vagal fibers that innervate the heart are primarily A-fibers, which are the largest and fastest conducting axons that originate from nerve cells located primarily in the nucleus ambiguus. The nucleus ambiguus also receives and integrates information from the cortical and subcortical systems described above.[118] Thus, the vagal regulatory centers respond to peripheral sensory (afferent) inputs and higher brain-center inputs to adjust neuronal outflows, which results in the vagally mediated beat-to-beat changes in HR.

Increased efferent activity in the vagal nerves (also called the 10th cranial nerve) slows HR and increases bronchial tone. The vagus nerves are the primary nerves for the parasympathetic system and they innervate the intrinsic cardiac nervous system. A few of these connections synapse on motor neurons in the intrinsic cardiac nervous system and these neurons project directly to the SA node (and other tissues in the heart), where they trigger acetylcholine release to slow HR.[11] However, the majority of the efferent preganglionic vagal neurons (~80%) connect to local circuitry neurons in the intrinsic cardiac nervous system, where motor information is integrated with inputs from mechanosensory and chemosensory neurons in the heart.[119] Thus, efferent sympathetic and parasympathetic activity is integrated in and with the activity occurring in the heart's intrinsic nervous system, including the input signals from the mechanosensory and chemosensory neurons within the heart, all of which ultimately contribute to beat-to-beat cardiac functional changes.[17]

In summary, the cardiorespiratory control system is complex and information from many inputs is integrated at multiple levels of the system, all of which are important for the generation of normal beat-to-beat variability in HR and BP. The medulla oblongata is the major structure integrating incoming afferent information from the heart, lungs and face with inputs from cortical and subcortical structures and is the source of the respiratory modulation of the activity patterns in sympathetic and parasympathetic outflow. The intrinsic cardiac nervous system integrates mechanosensitive and chemosensitive neuron inputs with efferent information from both the sympathetic and parasympathetic inputs from the brain, and as a complete system affects HRV, vasoconstriction and cardiac contractility in order to regulate HR and blood pressure.[120]

HRV and Analysis Methods

The normal variability in heart rate results from the descending (efferent) and the ascending (afferent) activity occurring in the two branches of the ANS, which act in concert, along with mechanical, hormonal and other physiological mechanisms to maintain cardiovascular parameters in their optimal ranges and to permit appropriate adjustments to changing external and internal conditions and challenges (Figure 1.3).

At rest, both sympathetic and parasympathetic nerves are tonically active, with the vagal effects predominant. Therefore, heart rate best reflects the relative balance between the sympathetic and parasympathetic systems. When speaking of autonomic balance, it should be kept in mind that a healthy system is constantly and dynamically changing. Therefore, an important indicator of the health status of the regulatory systems is that they have the capacity to respond to and adjust the relative autonomic balance, as reflected in heart rate, to the appropriate state for whatever a person is engaged in at any given moment. In other words, does the HR dynamically respond and is it higher in the daytime or when dealing with challenging tasks and lower when at rest or during sleep? Inability of the physiological self-regulatory systems to adapt to the current context and situation is associated with numerous clinical conditions.[121] Also, distinct, altered, circadian patterns in 24-hour heart rates are associated with different and specific psychiatric disorders, particularly during sleep.[122, 123]

Heart rate estimated at any given time represents the net effect of the neural output of the parasympathetic (vagus) nerves, which slows HR and the sympathetic nerves, which accelerate it. In a denervated human heart in which there are no connections from the ANS to the heart following its transplantation, the intrinsic rate generated by a pacemaker (SA node) is about 100 BPM.[124] Parasympathetic activity predominates when HR is below this intrinsic rate during normal daily activities and when at rest or sleep. When HR is above ~100 BPM, the relative balance shifts and sympathetic activity predominates. The average 24-hour HR in healthy people is ~73 BPM. Higher HRs are independent markers of mortality in a wide spectrum of conditions.[121]

It is important to note the natural relationship between HR and amount of HRV. As HR increases there is less time between heartbeats for variability to occur, so HRV decreases, while at lower HRs there is more time between heartbeats, so variability naturally increases. This is called *cycle length dependence*, and it persists in the healthy elderly to a variable degree, even at very advanced ages. However, elderly patients with ischemic heart disease or other pathologies increasingly have less variability as HRs decrease, ultimately losing the relationship between HR and variability – to the point that variability does not increase at all with reductions in HR.[125] Even in healthy subjects, the effects of cycle length dependence should be taken into account when assessing HRV, and HR values should always be reported, especially when HRs are increased because of factors such as stress reactions, medications and physical activity.

An increase in sympathetic activity is the principal method used to increase HR above the intrinsic level generated by the SA node. Activation of this branch of the ANS, in concert with the activation of the endocrine system, facilitates the ability to respond to challenges, stressors or threats by increasing the mobilization of energy resources.

Following the onset of sympathetic stimulation, there is a delay of up to 5 seconds before the stimulation induces a progressive increase in HR, which reaches a steady level in 20 to 30 seconds if the stimulus is continuous.[120] The relatively slow response to sympathetic stimulation is in direct contrast to vagal stimulation, which is almost instantaneous. However, the effect of sympathetic stimulation on HR is longer-lasting and even a brief stimulus can affect HR for 5 to 10 seconds. Efferent (descending) sympathetic nerves target the SA node via the intrinsic cardiac nervous system and the bulk of the myocardium (heart muscle). Action potentials conducted by these motor neurons trigger norepinephrine and epinephrine release, which increases HR and strengthens the contractility of the atria and ventricles.

HRV can be assessed with various analytical approaches, although the most commonly used are frequency domain (power spectral density) analysis and time domain analysis. In both methods, the time intervals between each successive normal QRS complex are first determined. All abnormal beats not generated by the sinus node are eliminated from the record. The interactions between autonomic neural activity, BP, respiratory and higher-level control systems produce both short- and long-term rhythms in HRV measurements.[5, 126, 127] The most common form for observing these changes is the heart-rate tachogram, a plot of the sequence of time intervals between heartbeats (Figure 3.3).

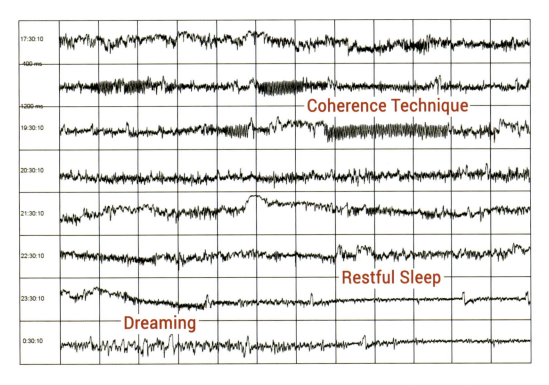

Figure 3.3. An example of the heart-rate tachogram, a plot of the sequence of time intervals between heartbeats over an 8-hour period in ambulatory recording taken from a 36-year-old male. Each of the traces is one hour long, with the starting time of the hour on the left-hand side of the figure. The time between each vertical line is 5 minutes. The vertical axis within each of the hourly tracings is the time between heartbeats (interbeat intervals) ranging from 400 tp 1,200 milliseconds (label shown on second row). A 15-minute period of HRV coherence can be seen in the latter part of the hour, starting at 19:30 when this man practiced HeartMath's Heart Lock-In® Technique. The latter part of the hour, starting at 23:30, is typical of restful sleep.

Power spectral analysis is used to separate the complex HRV waveform into its component rhythms (Figure 3.4). Spectral analysis provides information about how power is distributed (the variance and amplitude of a given rhythm) as a function of frequency (the time period of a given rhythm). The main advantages of spectral analysis over the time domain measures are that it supplies both frequency and amplitude information on the specific rhythms that exist in the HRV waveform, providing a means to quantify these oscillations over any given period. The values are expressed as the power spectral density, which is the area under the curve (peak) in a given bandwidth of the spectrum. The power or height of the peak at any given frequency indicates the amplitude and stability of the rhythm. The frequency reflects the period of time over which the rhythm occurs. For example, a 0.1 hertz frequency has a period of 10 seconds. In order to understand how power spectral analysis distinguishes the various underlying physiological mechanisms reflected in the heart's rhythm, a brief discussion of the underlying physiological mechanisms is helpful.

The power spectrum is divided into three main frequency ranges.

High-Frequency Band

The high-frequency (HF) spectrum is the power in the range from 0.15 to 0.4 hertz, which equates to rhythms with periods that occur between 2.5 and 7 seconds. This band reflects parasympathetic or vagal activity and is frequently called the respiratory band because it corresponds to the HR variations related to the respiratory cycle known as respiratory sinus arrhythmia. The mechanisms linking the variability of HR to respiration are complex and involve both central and reflex interactions.[118] During inhalation, the cardiorespiratory center inhibits vagal outflow, resulting in speeding up HR. Conversely, during exhalation, vagal outflow is restored, resulting in slowing HR.[128] The magnitude of the oscillation is variable, but in healthy people, it can be increased by slow, deep breathing.

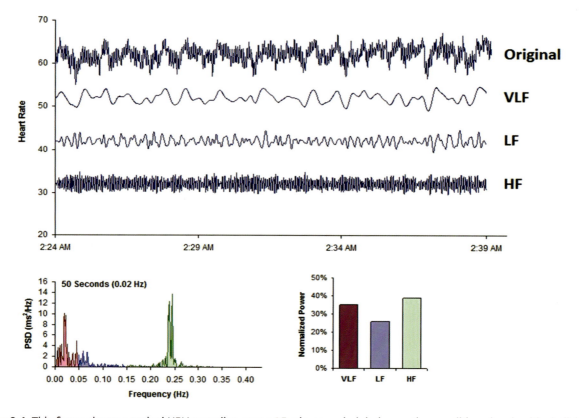

Figure 3.4. This figure shows a typical HRV recording over a 15-minute period during resting conditions in a healthy individual. The top trace shows the original HRV waveform. Filtering techniques were used to separate the original waveform into VLF, LF, and HF bands as shown in the lower traces. The bottom of the figure shows the power spectra (left) and the percentage of power (right) in each band.

Reduced parasympathetic (HF) activity has been found in a number of cardiac pathologies as discussed earlier. In terms of psychological regulation, reduced vagally mediated HRV has been linked to reduced self-regulatory capacity and cognitive functions that involve the executive centers of the prefrontal cortex. This is consistent with the finding that lower HF power is associated with stress, panic and anxiety/worry. Lower parasympathetic activity, rather than reduced sympathetic functioning, appears to account for a higher ratio of the reduced HRV in aging.[96]

Low-Frequency Band

The low-frequency (LF) band ranges between 0.04 and 0.15 hertz, which equates to rhythms or modulations with periods that occur between 7 and 25 seconds. This region was previously called the baroreceptor range or midfrequency band by many researchers because it primarily reflects baroreceptor activity while at rest.[129] As discussed previously, the vagus nerves are a major conduit through which afferent neurological signals from the heart are relayed to the brain, including baroreflex signals. Baroreceptors are stretch-sensitive mechanoreceptors located in the chambers of the heart and vena cavae, carotid sinuses (which contain the most sensitive mechanoreceptors) and the aortic arch. Baroreflex gain is commonly calculated as the beat-to-beat change in HR per unit of change in BP. Decreased baroreflex gain is related to aging and impaired regulatory capacity.

The existence of a cardiovascular system resonance frequency, which is caused by the delay in the feedback loops in the baroreflex system, has long been established. When the cardiovascular system oscillates at this frequency, there is a distinctive high-amplitude peak in the HRV power spectrum around 0.1 hertz. Most mathematical models show that the resonance frequency of the human cardiovascular system is determined by the feedback loops between the heart and brain.[130, 131] In humans and many other mammals, the resonance frequency of the system is approximately 0.1 hertz, equivalent to a 10-second rhythm, which is also characteristic of the coherent state described earlier.

The sympathetic nervous system does not appear to have much influence in rhythms above 0.1 hertz, while the parasympathetic system can be observed to affect heart rhythms down to 0.05 hertz (20-second rhythm). Therefore, during periods of slow respiration rates, vagal activity can easily generate oscillations in the heart rhythms that cross over into the LF band.[111, 132, 133] Thus, respiratory-related efferent vagally mediated influences are particularly present in the LF band when respiration rates are below 8.5 breaths per minute/7-second periods or when an individual sighs or takes a deep breath.[133, 134]

In ambulatory 24-hour HRV recordings, it has been suggested that the LF band reflects sympathetic activity and the LF/HF ratio has been used, controversially so, to assess the balance between sympathetic and parasympathetic activity.[135-137] A number of researchers have challenged this perspective and have persuasively argued that in resting conditions, the LF band reflects baroreflex activity and not cardiac sympathetic innervation.[40, 71, 96, 105-107]

The perspective that the LF band reflects sympathetic activity comes from observations of 24-hour ambulatory recordings in which there are frequent sympathetic activations primarily resulting from physical activity, but also emotional reactions, which can create oscillations in the heart rhythms that cross over from the VLF band into the lower region of the LF band. In long-term ambulatory recordings, the LF band fairly approximates sympathetic activity when increased sympathetic activity occurs.[138] Unfortunately, some authors have assumed that this interpretation also is true of short-term resting recordings and have confused slower breathing-related increases in LF power with sympathetic activity, when in reality it is almost entirely vagally mediated.

Very-Low-Frequency Band

The very-low-frequency band (VLF) is the power in the HRV power spectrum range between 0.0033 and 0.04 hertz which equates to rhythms or modulations with periods that occur between 25 and 300 seconds. Although all 24-hour clinical measures of HRV reflecting low HRV are linked with increased risk of adverse outcomes, the VLF band has stronger associations with all-cause mortality than LF and HF bands.[98, 139-141] Low VLF power has been shown to be associated with arrhythmic death[142] and PTSD.[143] Additionally, low power in this band has been associated with high inflammation[144, 145] in a number of studies and has been correlated with low levels of testosterone, while other biochemical markers, such as those mediated by the HPA axis (e.g., cortisol), have not.[146] Longer time periods using 24-hour HRV recordings should be obtained to provide comprehensive assessment of VLF and ULF fluctuations.[147]

Historically, the physiological explanation and mechanisms involved in the generation of the VLF component have not been as well defined as the LF and HF components. This region has been largely ignored even though it is the most predictive of adverse outcomes. Long-term regulation mechanisms and ANS activity related to thermoregulation, the renin-angiotensin system and other hormonal factors appear to contribute to this band.[148, 149] Recent work by Dr. J. Andrew Armour has shed new light on the mechanisms underlying the VLF rhythm and suggests that we have to reconsider both the mechanisms and importance of this band.

This line of research began after some surprising results from a study looking at HRV in autotransplanted hearts in dogs. In autotransplants, the heart is removed and placed back in the same animal, so there is no need for anti-rejection medications. The primary purpose of the study was to determine if the autonomic nerves reinnervated the heart post-transplant. Monthly 24-hour HRV recordings were done over a one-year period on all of the dogs with autotransplanted hearts as well as control dogs. It turned out that the nerves did reinnervate, but in a way that was not accurately reflected in HRV. It was shown that the intrinsic cardiac nervous system had neuroplasticity and restructured its neural connections. The truly surprising result was that these de-innervated hearts had higher levels of HRV than the control dogs immediately post-transplant and these levels were sustained over a one-year period, including HRV, which typically is associated with respiration (Figure 3.5).[150] This was unexpected because in human transplant recipients, there is very little HRV.[151]

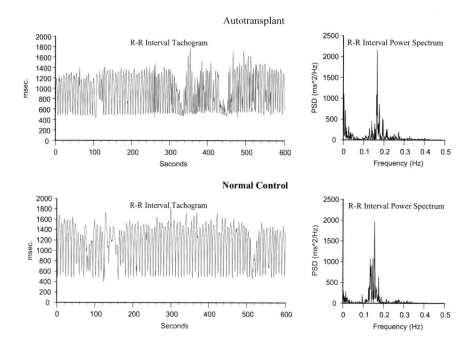

Figure 3.5. Heart Rhythms Generated by a Transplanted Heart: At top left is the heart-rate tachogram of a dog after undergoing cardiac autotransplantation, with the accompanying top-right graph showing the HRV power spectrum. For comparison, the bottom graphs show the heart-rate tachogram and HRV power spectrum of a normal dog. Note the similarity between the two.

Following up on these results, Armour and colleagues developed methods for obtaining long-term single-neuron recordings from a beating heart and, simultaneously, from extrinsic cardiac neurons.[13] This work, combined with later findings by Kember and Armour, implies that the VLF rhythm is generated by the stimulation of afferent sensory neurons in the heart, which in turn activates various levels of the feedback and feed-forward loops in the heart's intrinsic cardiac nervous system, as well as between the heart and neurons in the extrinsic cardiac ganglia and spinal column.[152, 153] Thus, the VLF rhythm appears to be produced by the heart itself and is an intrinsic rhythm that appears to be fundamental to health and well-being. Armour has observed that when the amplitude of the VLF rhythm at the neural level is diminished in an animal research subject, the animal is in danger and will expire shortly if the research procedures proceed. This cardiac origin of the VLF rhythm also is supported by studies showing that sympathetic blockade does not affect VLF power and VLF activity remains in tetraplegics, whose sympathetic innervation of the heart and lungs is disrupted.[154]

Circadian rhythms, core body temperature, metabolism, hormones and intrinsic rhythms generated by the heart all contribute to lower-frequency rhythms (e.g., very-low-frequency and ultra-low-frequency rhythms) that extend below 0.04 hertz. In healthy individuals, there is an increase in VLF power that occurs during the night and peaks before waking.[155,156] This increase in autonomic activity appears to correlate with the morning cortisol peak.

To summarize, experimental evidence suggests the VLF rhythm is intrinsically generated by the heart and the amplitude and frequency of these oscillations are modulated by efferent sympathetic activity. Normal VLF power appears to indicate healthy function, and increases in resting VLF power and/or shifting of frequency can reflect efferent sympathetic activity. The modulation of the frequency of this rhythm resulting from physical activity,[157] stress responses and other factors that increase efferent sympathetic activation can cause it to cross over into the lower region of the LF band during ambulatory monitoring or during short-term recordings when there is a significant emotional stressor.[5]

Time Domain Measurements of HRV

Time domain indices quantify the amount of variance in the interbeat interval (IBI) using statistical measures. Time domain measures are the simplest to calculate. Time domain measures do not provide a means to adequately quantify autonomic dynamics or determine the rhythmic or oscillatory activity generated by the different physiological control systems. However, since they are always calculated the same way, data collected by different researchers are comparable, but only if the recordings are exactly the same length of time and the data are collected under the same conditions. The three most important and commonly reported time domain measures are the SDNN, the SDNN index, and the RMSSD.

SDNN

The SDNN is the standard deviation of the normal-to-normal (NN) sinus-initiated interbeat-intervals measured in milliseconds. This measure reflects the ebb and flow of all the factors that contribute to HRV. In 24-hour recordings, the SDNN is highly correlated with ULF and total power.[96] In short-term resting recordings, the primary source of the variation is parasympathetically mediated, especially with slow, deep-breathing protocols. However, in ambulatory and longer-term recordings the SDNN values are highly correlated with lower-frequency rhythms.[83] Thus, low age-adjusted values predict morbidity and mortality. For example, patients with moderate SDNN values (50-100 milliseconds) have a 400% lower risk of mortality than those with low values (0-50 milliseconds) in 24-hour recordings.[158, 159]

SDNN Index

The SDNN index is the mean of the standard deviations of all the NN intervals for each 5-minute segment. Therefore, this measurement only estimates variability due to the factors affecting HRV within a 5-minute period. In 24-hour HRV recordings, it is calculated by first dividing the 24-hour record into 288 five-minute segments and then calculating the standard deviation of all NN intervals contained within each segment. The SDNN Index is the average of these 288 values.[90] The SDNN index is believed to primarily measure autonomic influence on HRV. This measure tends to correlate with VLF power over a 24-hour period.[83]

RMSSD

The RMSSD is the root mean square of successive differences between normal heartbeats. This value is obtained by first calculating each successive time difference between heartbeats in milliseconds. Each of the values is then squared and the result is averaged before the square root of the total is obtained. The RMSSD reflects the beat-to-beat variance in heart rate and is the primary time domain measure used to estimate the vagally mediated changes reflected in HRV.[90] The RMSSD is correlated with HF power and therefore also reflects self-regulatory capacity, as discussed earlier.[83]

HRV Assessment Services

The *Autonomic Assessment Report*, (AAR), developed by the HeartMath Research Center, provides physicians, researchers and mental health-care professionals with a diagnostic tool to detect abnormalities and imbalances in the autonomic nervous system and predict those at increased risk of developing various pathologies often before symptoms become manifest. The HeartMath Research Center provides this analysis service to physicians and medical institutions throughout the U.S. and abroad.

The Autonomic Assessment Report is a powerful tool for quantifying autonomic function. The AAR provides health-care professionals and researchers with a non-invasive test that quantifies autonomic function and relative balance and risk stratification, and assesses the effects of interventions on autonomic function. The AAR is derived from 24-hour ambulatory ECG recordings, typically obtained with an "HRV" recorder, which is inexpensive, lightweight and comfortable to wear. The AAR is based on analysis of heart rate variability, which provides a unique window into the interactions of sympathetic and parasympathetic control of the heart. The report includes time domain, frequency domain and circadian rhythm analysis, which together constitute a comprehensive analysis of autonomic activity, relative balance and rhythms. Time domain measures include the mean normal-to-normal (NN) intervals during a 24-hour recording and statistical measures of the variance between NN intervals. Power spectral density analysis is used to assess how power is distributed as a function of frequency, providing a means to quantify autonomic balance at any given point in the 24-hour period, as well as to chart the circadian rhythms of activity in the two branches of the autonomic nervous system. HMI has established and maintains an extensive HRV database of healthy individuals that greatly increases the AAR's value as a diagnostic and risk-assessment tool. Additionally, the age and gender normative values are provided for each time and frequency domain HRV value.

HRV is useful for monitoring autonomic function and assessing ANS involvement in a number of clinical conditions. Importantly, low HRV has been found to be predictive of increased risk of heart disease, sudden cardiac death as well as all-cause mortality.

Autonomic Function Imbalances Are Associated With:

- Depression
- Hypoglycemia
- Panic Disorder
- Sleep Disorder
- Asthma
- Fatigue
- Dizziness
- Nausea
- Irritable Bowel
- Fibromyalgia
- Hypertension
- Chemical Sensitivity
- Premenstrual Syndrome
- Anxiety
- Migraine
- Arrhythmia

Autonomic imbalances have been implicated in a wide variety of pathologies, including depression, fatigue, premenstrual syndrome, hypertension, diabetes mellitus, ischemic heart disease, coronary heart disease and environmental sensitivity. Stress and emotional states have been shown to dramatically affect autonomic function. Self-regulation techniques, which enable individuals to gain greater control of their mental and emotional stress and improve their autonomic

Chapter 3: HRV: An Indicator of Self-Regulatory Capacity, Autonomic Function and Health

functioning, can significantly affect a wide variety of disorders in which autonomic imbalance plays a role. The AAR analysis is highly useful for the quantitative demonstration of the effects of HeartMath interventions in restoring healthy autonomic function in many patients who have been able to significantly improve their symptomatology and psychological well-being through practice of these techniques.

The Autonomic Assessment Report Interpretation Guide and Instructions booklet, available from HMI, provides clinicians with understandable descriptions of HRV measures used in the report and how to interpret them in clinical applications. It includes a number of case histories and clinical examples.

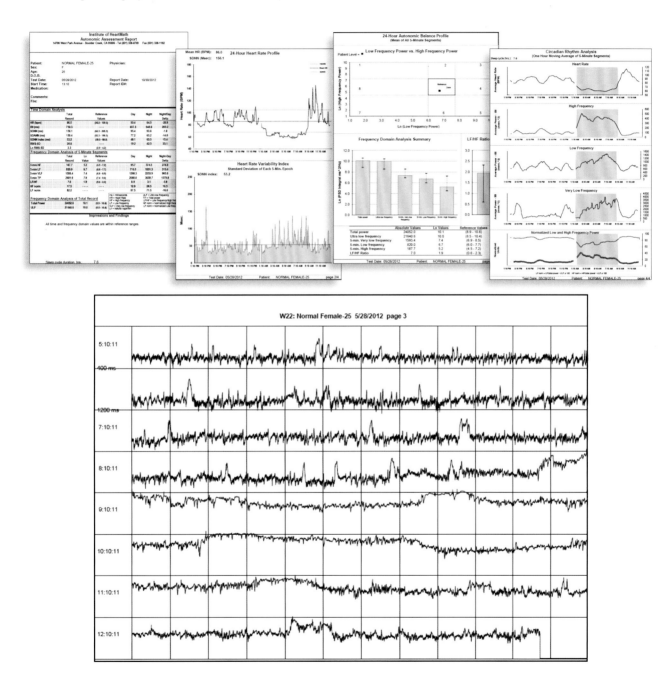

Figure 3.6. Sample pages from the HeartMath Autonomic Assessment Report. Shown from top left to right are: (1) Summary page with normative reference ranges. (2) 24-Hour Heart Rate Profile and Heart Rate Variability Index plot. (3) Autonomic Balance Profile and frequency domain analysis summary. (4) Circadian Rhythm Analysis page and bottom graph. (5) One page of the three heart-rate tachogram pages showing HRV from the full 24-hours.

CHAPTER 4

Coherence

> **Definitions of Coherence**
> **Clarity of thought, speech and emotional composure**
> *The quality of being orderly, consistent and intelligible (e.g. a coherent sentence).*
> **Synchronization or entrainment between multiple waveforms**
> *A constructive waveform produced by two or more waves that are phase- or frequency-locked.*
> **Order within a singular oscillatory waveform**
> *An ordered or constructive distribution of power content within a single waveform; autocoherence (e.g. sine wave).*

Many contemporary scientists believe it is the underlying state of our physiological processes that determines the quality and stability of the feelings and emotions we experience. The feelings we label as positive actually reflect body states that are coherent, meaning "the regulation of life processes becomes efficient, or even optimal, free-flowing and easy,"[160] and the feelings we label as "negative," such as anger, anxiety and frustration are examples of incoherent states. It is important to note, however, these associations are not merely metaphorical. For the brain and nervous system to function optimally, the neural activity, which encodes and distributes information, must be stable and function in a coordinated and balanced manner. The various centers within the brain also must be able to dynamically synchronize their activity in order for information to be smoothly processed and perceived. Thus, the concept of coherence is vitally important for understanding optimal function.

The various concepts and measurements embraced under the term coherence have become central to fields as diverse as quantum physics, cosmology, physiology and brain and consciousness research.[59] Coherence has several related definitions, all of which are applicable to the study of human physiology, social interactions and global affairs. The most common dictionary definition is the quality of being logically integrated, consistent and intelligible, as in a coherent statement.[159] A related meaning is the logical, orderly and aesthetically consistent relationship among parts.[159] Coherence always implies correlations, connectedness, consistency and efficient energy utilization. Thus, coherence refers to wholeness and global order, where the whole is greater than the sum of its individual parts.

In physics, coherence also is used to describe the coupling and degree of synchronization between different oscillating systems. In some cases, when two or more oscillatory systems operate at the same basic frequency, they can become either phase- or frequency-locked, as occurs between the photons in a laser.[160] This type of coherence is called *cross-coherence* and is the type of coherence that most scientists think of when they use the term. In physiology, cross-coherence occurs when two or more of the body's oscillatory systems, such as respiration and heart rhythms, become entrained and operate at the same frequency.

Another aspect of coherence relates to the dynamic rhythms produced by a single oscillatory system. The term *autocoherence* describes coherent activity within a single system. An ideal example is a system that exhibits sine-wavelike oscillations; the more stable the frequency, amplitude and shape, the higher the degree of coherence. When coherence is increased in a system that is coupled to other systems, it can pull the other systems into increased synchronization and more efficient function.

For example, frequency pulling and entrainment can easily be seen between the heart, respiratory and blood-pressure rhythms as well as between very-low-frequency brain rhythms, craniosacral rhythms and electrical potentials measured across the skin.[142, 143]

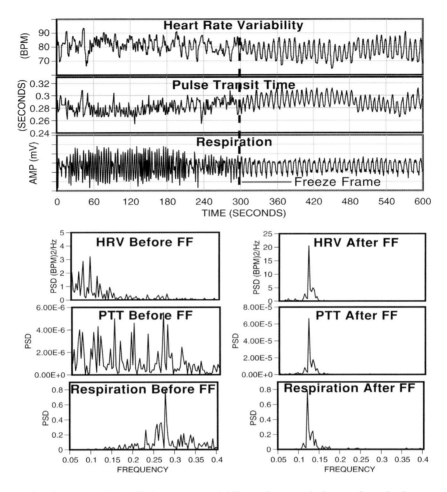

Figure 4.1. The top graphs show an individual's heart rate variability, pulse transit time and respiration patterns for 10 minutes. At the 300-second mark, the individual did HeartMath's Freeze Frame Technique and all three systems came into entrainment, meaning the patterns were harmonious instead of scattered and out of sync. The bottom graphs show the spectrum analysis view of the same data. The left-hand side is the spectral analysis before Freeze-Framing. Notice how each pattern looks quite different from the others. The graphs on the right show how all three systems are entrained at the same frequency after Freeze-Framing.

Global Coherence

For any system to produce a meaningful function, it must have the property of global coherence. In humans, this includes our physical, mental, emotional and social systems. However, the energy efficiency and degree of coordinated action of any given system can vary widely and does not necessarily result in a coherent output or flow of behavior. Global coherence does not mean everyone or all parts of a system are doing the same thing simultaneously. In complex globally coherent systems, such as human beings, there is a vast amount of activity at every level of magnification or scale that spans more than two-thirds of the 73 known octaves of the electromagnetic spectrum.[165] It can appear at one level of scale that a given system is operating autonomously, yet it is perfectly coordinated within the whole. In living systems, there are microlevel systems, molecular machines, protons and electrons, organs and glands, each functioning autonomously, doing very different things at different rates, yet all working together in a complex harmoniously coordinated and synchronized manner. If this were not happening, it would be a free-for-all among the body's independent systems, rather than a coordinated federation of interdependent systems and functions. Biologist Mae-Won Ho has suggested that coherence is the defining quality of living systems and accounts for their most characteristic properties, such as long-range order and coordination, rapid and efficient energy transfer and extreme sensitivity to specific signals.[165]

We introduced the term *physiological coherence* to describe the degree of order, harmony and stability in the various rhythmic activities within living systems over any given time period.[163] This harmonious order signifies a coherent system, whose efficient or optimal function is directly related to the ease and flow in life processes. In contrast, an erratic, discordant pattern of activity denotes an incoherent system whose function reflects stress and inefficient utilization of energy in life processes. Specifically, heart coherence (also referred to as cardiac coherence or resonance) can be measured by HRV analysis wherein a person's heart-rhythm pattern becomes more ordered and sine wave-like at a frequency of around 0.1 hertz (10 seconds).

When a person is in a more coherent state there is a shift in the relative autonomic balance toward increased parasympathetic activity (vagal tone), increased heart-brain synchronization and entrainment between diverse physiological systems. In this mode, the body's systems function with a high degree of efficiency and harmony and natural regenerative processes are facilitated. Although physiological coherence is a natural human state that can occur spontaneously, sustained episodes generally are rare. While some rhythmic-breathing methods may induce coherence for brief periods, our research indicates that people can achieve extended periods of physiological coherence by actively self-generating positive emotions.

When functioning in a coherent mode, the heart pulls other biological oscillators into synchronization with its rhythms, thus leading to entrainment of these systems (Figure 4.1). Entrainment is an example of a physiological state in which there is increased coherence *between* multiple oscillating systems and also *within* each system. Thus, our findings essentially underscore what people have intuitively known for some time: Positive emotions not only "feel better," they actually tend to increase synchronization of the body's systems, thereby enhancing energy and enabling us to function with greater efficiency and effectiveness.

The coherence model takes a dynamic systems approach that focuses on increasing people's self-regulatory capacity through self-management techniques that induce a physiological shift, which is reflected in the heart's rhythms. We also suggest that rhythmic activity in living systems reflects the regulation of interconnected biological, social and environmental networks and that important biologically relevant information is encoded in the dynamic patterns of physiological activity. For example, information is encoded in the time interval between action potentials in the nervous system and patterns in the pulsatile release of hormones. Our research also suggests that the time intervals between heartbeats (HRV) also encode information, which is communicated across multiple systems and helps synchronize the system as whole. The afferent pathways from the heart and blood vessels are given more relevance in this model because of the significant degree of afferent cardiovascular input to the brain and the consistent generation of dynamic patterns generated by the heart. Our perspective is that positive emotions in general, including self-induced positive emotions, shift the entire system into a more globally coherent and harmonious physiological mode, one that is associated with improved system performance, ability to self-regulate and overall well-being. The coherence model predicts that different emotions are reflected in state-specific patterns in the heart's rhythms[5] independent of the amount of HRV/HR (Figure 4.2). Recent independent work has verified this by demonstrating a 75% accuracy rate in detection of discrete emotional states from the HRV signal using a neural network approach for pattern recognition.[164] In a study of the effects of playing violent and nonviolent video games, it was found that when playing violent video games, the players had lower cardiac coherence levels and higher aggression levels than did nonviolent game players and that higher levels of coherence were negatively related to aggression.[165]

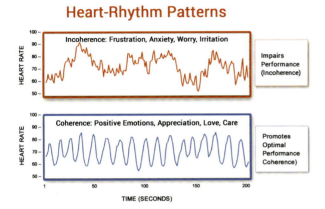

Figure 4.2. Heart-rhythm patterns.

The coherent state has been correlated with a general sense of well-being and improvements in cognitive, social and physical performance. We have observed this association between emotions and heart-rhythm patterns in studies conducted in both laboratory and natural settings and for both spontaneous and intentionally generated emotions.[163, 168]

Several studies in healthy subjects, which helped inform the model, show that during the experience of positive emotions, a sine-wavelike pattern naturally emerges in the heart's rhythms without any conscious changes in breathing.[51, 133] This is likely because of more organized outputs of the subcortical structures involved in processing emotional information, as described by Pribram,[169] Porges,[82] Oppenheimer and Hopkins[117] and Thayer,[89] in which the subcortical structures influence the oscillatory output of the cardiorespiratory control system in the medulla oblongata.

A brief summary of the psychophysiological coherence model is provided below. A detailed discussion on the nature of coherence can be found in two seminal articles.[5, 59]

The Coherence Model Postulates:

1. The functional status of the underlying psychophysiological system determines the range of one's ability to adapt to challenges, self-regulate and engage in harmonious social relationships. Healthy physiological variability, feedback systems and inhibition are key elements of the complex system for maintaining stability and capacity to appropriately respond to and adapt to changing environments and social demands.

2. The oscillatory activity in the heart's rhythms reflects the status of a network of flexible relationships among dynamic interconnected neural structures in the central and autonomic nervous systems.

3. State-specific emotions are reflected in the patterns of the heart's rhythms independent of changes in the amount of heart rate variability.

4. Subcortical structures constantly compare information from internal and external sensory systems via a match/mismatch process that evaluates current inputs against past experience to appraise the environment for risk or comfort and safety.

5. Physiological or cardiac coherence is reflected in a more ordered sine-wavelike heart-rhythm pattern associated with increased vagally mediated HRV, entrainment between respiration, blood pressure and heart rhythms and increased synchronization between various rhythms in the EEG and cardiac cycle.

6. Vagally mediated efferent HRV provides an index of the cognitive and emotional resources needed for efficient functioning in challenging environments in which delayed responding and behavioral inhibition are critical.

7. Information is encoded in the time between intervals (action potentials, pulsatile release of hormones, etc.). The information contained in the interbeat intervals in the heart's activity is communicated across multiple systems and helps synchronize the system as a whole.

8. Patterns in the activity of cardiovascular afferent neuronal traffic can significantly influence cognitive performance, emotional experience and self-regulatory capacity via inputs to the thalamus, amygdala and other subcortical structures.

9. Increased "rate of change" in cardiac sensory neurons (transducing BP, rhythm, etc.) during coherent states increases vagal afferent neuronal traffic, which inhibits thalamic pain pathways at the level of the spinal cord.

10. Self-induced positive emotions can shift psychophysiological systems into more globally coherent and harmonious orders that are associated with improved performance and overall well-being.

The coherence model includes specific approaches for quantifying the various types of physiological coherence measures, such as cross-coherence (frequency entrainment between respiration, BP and heart rhythms), or synchronization among systems (e.g., synchronization between various EEG rhythms and the cardiac cycle), autocoherence (stability of a single waveform such as respiration or HRV patterns) and system resonance.[5] A coherent heart rhythm is defined as a relatively harmonic, sine-wavelike signal with a very narrow, high-amplitude peak in the low-frequency (LF) region of the HRV power spectrum with no major peaks in the very-low-frequency (VLF) or high-frequency (HF) regions. Physiological coherence is assessed by identifying the maximum peak in the 0.04 to 0.26 hertz range of the HRV power spectrum, calculating the integral in a window 0.030 hertz wide, centered on the highest peak in that region and then calculating the total power of the entire spectrum. The coherence ratio is formulated as (peak power/[total power − peak power]).[5]

Physiological Coherence

A state characterized by:

- High heart-rhythm coherence (sine-wavelike rhythmic pattern).
- Increased parasympathetic activity.
- Increased entrainment and synchronization between physiological systems.
- Efficient and harmonious functioning of the cardiovascular, nervous, hormonal and immune systems.

Social Coherence

Social coherence relates to pairs, family units, groups or larger organizations in which a network of relationships exists among individuals who share common interests and objectives. Social coherence is reflected as a stable, harmonious alignment of relationships that allow for the efficient flow and utilization of energy and communication required for optimal collective cohesion and action. There are, of course, cycles and variations in the quality of family, team or group coherence, similar to variations in an individual's coherence level. Coherence requires that group members are attuned and emotionally aligned and that the group's energy is globally organized and regulated by the group as a whole. Group coherence involves the same principles of global coherence described earlier, but in this context it refers to the synchronized and harmonious order in the relationships between and among the individuals rather than the systems within the body. The principles, however, remain the same: In a coherent team, there is freedom for the individual members to do their part and thrive while maintaining cohesion and resonance within the group's intent and goals. Anyone who has watched a championship sports team or experienced an exceptional concert knows that something special can happen in groups that transcends their normal performance. It seems as though the players are in sync and communicating on an unseen energetic level. A growing body of evidence suggests that an energetic field is formed between individuals in groups through which communication among all the group members occurs simultaneously. In other words, there is a literal group "field" that connects all the members. Sociologist Raymond Bradley, in collaboration with eminent brain researcher, neurosurgeon and neuroscientist Dr. Karl Pribram, developed a general theory of social communication to explain the patterns of social organization common to most groups and independent of size, culture, degree of formal organization, length of existence or member characteristics. They found that most groups have a global organization and coherent network of emotional energetic relations interconnecting virtually all members into a single multilevel hierarchy.[170]

CHAPTER 5

Establishing a New Baseline

At the HMI Research Center, we have found that the heart plays a central role in the generation of emotional experience and therefore, in the establishment of psychophysiological coherence. From a systems perspective, the human organism is truly a vast, multidimensional information network of communicating subsystems in which mental processes, emotions and physiological systems are inextricably intertwined. Whereas our perceptions and emotions were once believed to be dictated entirely by the brain's responses to stimuli arising from our external environment, the emerging perspectives in neuroscience more accurately describe perceptual and emotional experience as the composite of stimuli the brain receives from the external environment and the internal sensations or feedback transmitted to the brain from the bodily organs and systems.[5, 79] Thus, the heart, brain, nervous, hormonal and immune systems must all be considered fundamental components of the dynamic, interactive information network that determines our ongoing emotional experience.

Extensive work by Pribram has helped advance the understanding of the emotional system. In Pribram's model, past experience builds within us a set of familiar patterns that are established and maintained in the neural networks. Inputs to the brain from both the external and internal environments contribute to the maintenance of these patterns.

Research has shown that the heart's afferent neurological signals directly affect activity in the amygdala and associated nuclei, an important emotional processing center in the brain.[118] The amygdala is the key brain center that coordinates behavioral, immunological and neuroendocrine responses to environmental threats. It also serves as the processing center for emotional memory within the brain. In assessing the external environment, the amygdala scans the inputs (visual, auditory, smell) for emotional content and signals and compares them with stored emotional memories. In this way, the amygdala makes instantaneous decisions about the familiarity of incoming sensory information and because of its extensive connections to the hypothalamus and other autonomic nervous system centers is able to "hijack" the neural pathways activating the autonomic nervous system and emotional response before the higher brain centers receive the sensory information.[171]

One of the functions of the amygdala is to organize which patterns become "familiar" to the brain. If the rhythm patterns generated by the heart are disordered and incoherent, especially in early life, the amygdala learns to expect disharmony as the familiar baseline and thus we feel "at home" with incoherence, which can affect learning, creativity and emotional balance. In other words we feel "comfortable" with internal incoherence, which in this case actually is discomfort. On the basis of what has become familiar to the amygdala, the frontal cortex mediates decisions as to what constitutes appropriate behavior in any given situation. Thus, subconscious emotional memories and associated physiological patterns underlie and affect our perceptions, emotional reactions, thought processes and behavior.

"Since emotional processes can work faster than the mind, it takes a power stronger than the mind to bend perception, override emotional circuitry, and provide us with intuitive feeling instead. It takes the power of the heart."

– Doc Childre, HeartMath Institute founder

From our current understanding of the elaborate feedback networks between the brain, heart and mental and emotional systems, it becomes clear that the age-old struggle between intellect and emotion will not be resolved by the mind gaining dominance over the emotions, but rather by increasing the harmonious balance between the mental and emotional systems – a synthesis that provides greater access to our full range of intelligence.

Within the body, many processes and interactions occurring at different functional levels provide constant rhythmic inputs with which the brain becomes familiar. These inputs range from the rhythmic activity of the heart and our facial expressions, to digestive, respiratory and reproductive rhythms, to the constant interplay of messenger molecules produced by the cells of our body.

These inputs to the brain, translated into neural and hormonal patterns, are continuously monitored by the brain and help organize our perception, feelings and behavior. Familiar input patterns from the external environment and from within the body ultimately are written into neural circuitry and form a stable backdrop, or reference pattern, against which current and new information or experiences are compared. According to this model, when an external or internal input is sufficiently different from the familiar reference pattern, this "mismatch" or departure from the familiar underlies the generation of emotions.

The background physiological patterns with which the brain and body grow familiar are created and reinforced through life experiences and the way we perceive the world. It's important to note that the established patterns may not necessarily be positive or healthy for a person. For example, someone living in an environment that continually triggers anger or feelings of fear is likely to become familiar with these feelings and their neural and hormonal correlates. In contrast, an individual whose experience is dominated by feelings of security, love and care likely will become familiar with the physiological patterns associated with these feelings.

In order to maintain stability and feelings of safety and comfort, we must be able to maintain a match between our current experience or "reality" and one of our previously established neural programs.[172] When we encounter a new experience or challenge, there can be a mismatch between the input patterns of the new experience and the lack of a familiar reference. Depending on the degree of mismatch, it requires either an internal adjustment (self-regulation) or an outward behavioral action to reestablish a match and feeling of comfort. When a mismatch is detected from either external or internal sensory systems, a change in activity in the central and autonomic nervous systems is produced. If the response is short-lived (one to three seconds), it is called arousal or an orienting reflex. If, however, the stimulus or event is recurrent, the brain eventually adapts and we habituate by updating the memories that serve as the reference. For example, people who live in a noisy city adapt to the ambient noise and eventually tune it out. Subsequent to this adaptation, it is only when they take a trip to the quiet countryside that the actual lack of noise seems strange and is quite noticeable. The mismatch between the familiar noisy background and the quiet setting leads to an arousal reaction that gets our attention. It is this departure from the familiar that gives rise to a signaling function that creates the experience of an emotion, alerting us to the current state of the mismatch.

In addition to the monitoring and control processes for regulation "in the here-and-now," there are also appraisal processes that determine the degree of consistency or inconsistency between a current situation and the projected future. Appraisals of future outcomes can be broadly divided into optimistic and pessimistic.[173] Appraisals that project an inability to successfully deal with a situation may result in feelings of fear and anxiety. In keeping with the recent research on attentional bias,[172] this appraisal might not be accurate because it could be the result of hypersensitivity to cues that resemble past traumatic experiences in the current situation. Alternately, an inaccurate appraisal can be caused by an instability in the neural systems, or a lack of experience or insight of how to effectively deal with the projected future situation.[173] Despite the lack of accuracy of the appraisal, the familiarity of the input can be sufficient to elicit a pessimistic response. This means we can easily get "stuck" in unhealthy emotional and behavioral patterns and lasting improvements in emotional experience or behaviors cannot be sustained in the absence of establishing a new set point for the baseline. If behavior change or improved affective states are desired, it is therefore critical to focus on strategies that help to establish a new internal reference. As we successfully navigate new situations or challenges, the positive experience updates our internal reference. In essence, we mature through this process as we learn to more effectively self-regulate our emotions and deal with new situations and challenges. It is through this process that we are able to develop a new, healthier internal baseline reference against which we match inputs so that our assessments of benign inputs are more accurate and result in feelings of safety and comfort rather than threat and anxiety.

In a study of high school students who practiced self-regulation techniques over a four-month period, their resting HRV was significantly increased and the pattern of the HRV was significantly more coherent (Figure 5.1). These improvements in resting HRV coherence were significantly correlated with higher test scores and improved behaviors, suggesting that the practice of the self-regulation skills induces a more coherent heart rhythm, reinforcing the association in the sub-subcortical regulatory systems involved in a match/mismatch process between more coherent and stable rhythms in cardiovascular afferent neuronal traffic and feelings we perceive as positive.[110] By reinforcing this natural coupling in the sub-subcortical regulatory systems, the self-activation of a positive feeling can automatically initiate an increase in cardiac coherence, while at the same time, a physiological shift resulting from heart-focused breathing can help facilitate the experience of a positive emotion.

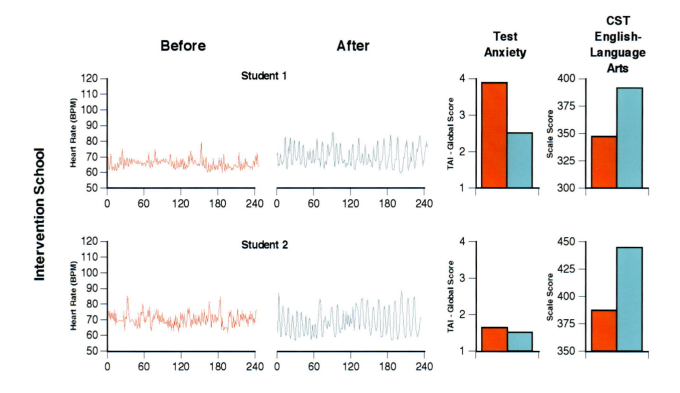

Figure 5.1 Typical resting state heart rate variability patterns in students. HRV recordings from the TestEdge National Demonstration Study showing examples of two students' resting-state heart-rhythm patterns, both before and approximately four months after the TestEdge intervention. Pre- and post-intervention test anxiety level (TAI-Global Scale score) and the CST–English Language Arts test score for each student are also shown. For the two students in the intervention school, the recordings show a shift from an erratic, irregular heart-rhythm pattern (left side), before the intervention, to a more coherent pattern (right side), which indicates the students had established a new more coherent baseline.[110]

Self-Regulation and Stability

Pribram and many others have conducted numerous experiments that provide evidence that the higher brain centers that monitor the pattern-matching process can self-regulate by inhibiting or "gating" the information flowing into the brain.[173] Where we focus our attention, for example, has a powerful effect on modulating inputs and thus on determining what gets processed at higher levels. In a noisy room filled with many conversations, for instance, we have the ability to tune out the noise and focus on a single conversation of interest. In a like manner, we can modulate pain from a stubbed toe or headache or desensitize ourselves to sensations like tickling and self-direct our emotions.[174] Ultimately, when we achieve control through the process of self-regulation, it results in feelings of satisfaction and gratification. In contrast, failure to effectively self-regulate and regain control often results in feelings of frustration, impatience, anxiety, overwhelm, hopelessness or depression.

If the neural systems that maintain the baseline reference patterns are unstable, unsettled emotions and atypical reactions are likely to be experienced. These neural systems can be destabilized by trauma, stress, anxiety or chemical stimulants, to name a few possibilities. Therefore, it is clear that responding in healthy and effective ways to ongoing inner and outer demands and circumstances, such as daily life situations, depends to a great extent on the synchronization, sensitivity and stability of our physiological systems.[5, 59]

Neural inputs originate from numerous organs and muscles, especially in the face. The heart and cardiovascular system, however, have far more afferent inputs than other organs and are the primary sources of consistent dynamic rhythms.[15] In addition to afferent nerve activity associated with mechanical information such as pressure and rate that occurs with each heartbeat, continuous dynamically changing patterns of afferent activity related to chemical information is sent to the brain and other systems in the body. In terms of emotional experience, there are afferent pathways to the amygdala via the nucleus of tractus solitarius and the activity in the central nucleus of the amygdala is synchronized to the cardiac cycle.[10, 177] Therefore, the afferent inputs from the cardiovascular system to the amygdala are important contributors in determining emotional experience and in establishing the set point to which the current inputs are compared.

In the context of this discussion, it is important to note that the heart's rhythmic patterns and the patterns of afferent neurological signals change to a more ordered and stable pattern when one uses HeartMath's heart-focused self-regulation techniques. Regular practice of these techniques, which include a shift of attentional focus to the center of the chest (heart area) accompanied by the conscious self-induction of a calm or positive emotional state, reinforces the association (pattern match) between a more coherent rhythm and a calm or positive emotion. Positive feelings then more automatically initiate an increase in cardiac coherence. Increased coherence initiated through heart-focused breathing tends to facilitate the felt experience of a positive emotion. Thus, practice affects the *repatterning process*. This is important in situations where there has been a sustained exposure to truly high-risk environments or trauma in the past, but which no longer are in effect and the patterns that developed in response to them no longer serve the individual in present safe environments.

Through this feed-forward process, regulatory capacity is increased and new reference patterns are established, which the system then strives to maintain, making it easier for people to maintain stability and self-directed control during daily activities, even in more challenging situations. Without a shift in the underlying baseline, it is exceedingly difficult to sustain behavioral change, placing people at risk of living their lives through the automatic filters of past familiar experience.

Self-Regulation Techniques that Reduce Stress and Enhance Human Performance

With stress levels continuing to rise around the world, people are becoming more conscious not only of the long-term effects of stress, but also of how unmanaged emotions compromise the quality of one's day-to-day life, limiting mental clarity, productivity, adaptability to life's challenges and enjoyment of its gifts.

"Failures of self-regulation are central to the vast majority of health and social problems that plague modern societies.
The most important strength that the majority of people need to build is the capacity to self-regulate their emotions, attitudes and behaviors."
– Rollin McCraty

It is commonly believed we have little control over the mind or emotions. For example, neuroscientist Joseph LeDoux, who studies brain circuits and the emotion of fear in animals, writes:

"Emotions are things that happen to us rather than things we will to occur. Although people set up situations to modulate their emotions all the time – going to movies and amusement parks, having a tasty meal, consuming alcohol and other recreational drugs – in these situations, external events are simply arranged so that the stimuli that automatically trigger emotions will be present. We have little direct control over our emotional reactions. Anyone who has tried to fake an emotion, or who has been the recipient of a faked one, knows all too well the futility of the attempt. While conscious control over emotions is weak, emotions can flood consciousness." [171], p. 19

While this is true for many people who have not developed their self-regulation skills, our research and experience show that the emotional system can be regulated and brought into coherence. This, of course, requires practice and effective skills, in much the same way that it takes techniques and practice to learn and develop mental or athletic skills.

The research on heart-brain interactions and intuition has informed the development of a set of self-regulation techniques and practices, the learning of which can be supported with the use of HRV coherence feedback technologies, collectively known as the HeartMath System.[178-182] The HeartMath System offers individuals a systematic and reliable means to intentionally self-regulate and shift out of a state of emotional unease or stress into a "new" positive state of emotional calm and stability. This occurs as a result of a practice in which an individual intentionally activates a positive or calm emotional state as a future target and activates a shift in patterns of the heart's activity to a more coherent state that enables the person to achieve and maintain stability and emotional composure.

The techniques are designed to enable people to intervene in the moment when negative and disruptive emotions are triggered, thus interrupting the body's normal stress response and initiating a shift toward increased coherence. This shift facilitates higher cognitive functioning, intuitive access and increased emotional regulation, all of which normally are compromised during stress and negative emotional states. The shift in the pattern of the heart's input to the brain thus serves to reinforce the self-generated positive emotional shift, making it easier to sustain. Through consistent use of HeartMath tools, the coupling between the psychophysiological coherence mode and positive emotions is further reinforced.

"The emotional frontier is truly the next frontier to conquer in human understanding.
The opportunity we face now, even before that frontier is fully explored and settled, is to develop our emotional potential and accelerate rather dramatically into a new state of being."
– Doc Childre

Self-Regulation Techniques That Increase Coherence

There is a paradigm shift emerging on behavioral intervention approaches that teach people self-regulation strategies that include a physiological aspect such as HRV biofeedback and that naturally increase vagal traffic. For example, there are many studies showing that the practice of breathing at 6 breaths per minute, supported by HRV biofeedback, induces the coherence rhythm and has a wide range of benefits.[183-189]

In addition to clinical applications, HRV coherence feedback training often is used to support self-regulation skill acquisition in educational, corporate, law enforcement and military settings. Several systems that assess the degree of coherence in the user's heart rhythms are available. The majority of these systems, such as emWave® Pro or Inner Balance® for iOS devices (HeartMath, Inc.), Relaxing Rhythms (Wild Divine) and the Stress Resilience Training System (Ease Interactive), use a noninvasive earlobe or finger pulse sensor and display the user's heart rhythm to provide feedback on the user's level of coherence.

Emotional self-regulation strategies may contribute to improved health and performance. Alone or in combination with HRV coherence biofeedback training, these strategies have been shown to increase resilience and accelerate recovery from stressors and trauma.[53, 58, 81, 190] Self-induced positive emotions can initiate a shift to increased cardiac coherence without any conscious intention to change the breathing rhythm.[51, 133] Typically, when people are able to self-activate a positive or calming feeling rather than remain focused on their breathing, they enjoy the shift in feeling and are able to sustain high levels of coherence for much longer time periods.[113]

Heart-focused self-regulation techniques and assistive technologies that provide real-time HRV coherence feedback provide a systematic process for self-regulating thoughts, emotions, behaviors and increasing physiological coherence. Many of these techniques (e.g., HeartMath's Heart-Focused Breathing, Freeze Frame, Inner Ease and Quick Coherence techniques[179] are designed to enable people to intervene in the moment they start to experience stress reactions or unproductive thoughts or emotions. With practice, one is able to use any of the techniques to shift into a more coherent physiological state before, during and after challenging or adverse situations, thus optimizing mental clarity, emotional composure and stability.

The first step in most of the techniques developed by the HeartMath Institute is called *Heart-Focused Breathing*, which includes placing one's attention in the center of the chest (the area of the heart) and imagining the breath is flowing in and out of the chest area while breathing a little slower and deeper than usual. Conscious regulation of one's respiration at a 10-second rhythm (five seconds in and five seconds out) (0.1 hertz) increases cardiac coherence and starts the process of shifting into a more coherent state.[5, 113] With conscious control over breathing, an individual can slow the rate and increase the depth of the breathing rhythm. This takes advantage of physiological mechanisms to modulate efferent vagal activity and thus the heart rhythm. This increases vagal afferent nerve traffic and increases the coherence (stability) in the patterns of vagal afferent nerve traffic. In turn, this influences the neural systems involved in regulating sympathetic outflow, informing emotional experience and synchronizing neural structures underlying cognitive processes.[5]

In addition to the self-regulation techniques that are primarily designed to be used in the moment, the Heart Lock-In Technique is more appropriate when one has more time to focus on sustaining a coherent state. It enables people to "lock in" the positive feeling states associated with the heart in order to boost their energy, heighten peace and clarity and effectively *retrain their physiology* to sustain longer periods of coherent function. With consistent practice, the Heart Lock-In facilitates the establishment of new reference patterns promoting increased physiological efficiency, mental acuity and emotional stability as a new baseline or norm.

While the HeartMath tools are intentionally designed to be easily learned and used in day-to-day life, our experience working with people of diverse ages, cultures, educational backgrounds and professions suggests that these techniques often facilitate profound shifts in perception, emotion and awareness. Moreover, extensive laboratory research performed at HMI has shown that the physiological changes accompanying such shifts are dramatic.

Several studies using various combinations of these self-regulation techniques have found significant correlations between HRV coherence and improvements in cognitive function and self-regulatory capacity.

For example:

> A study of middle school students with attention-deficit hyperactivity disorder showed a wide range of significant improvements in short- and long-term memory, ability to focus and significant improvements in behaviors both at home and in school.[108]

> A study of 41 fighter pilots engaging in flight simulator tasks found a significant correlation between higher levels of performance and heart-rhythm coherence as well as lower levels of frustration.[189]

> A study of recently returning soldiers from Iraq who were diagnosed with PTSD found that relatively brief periods of HRV coherence training combined with practicing the Quick Coherence Technique resulted in significant improvements in the ability to self-regulate along with a wide range of cognitive functions. The degree of improvement correlated with increased cardiac coherence.[109]

> Other studies have shown increases in parasympathetic activity (vagal tone),[133] reductions in cortisol and increases in DHEA,[116] decreases in blood pressure and stress measures in hypertensive populations,[113, 115] reduced health-care costs[112] and significant improvements in the functional capacity of patients with congestive heart failure.[192]

> A study of correctional officers showed reductions in systolic and diastolic BP, total cholesterol, fasting glucose, overall stress, anger, fatigue and hostility.[114] Similar results were obtained in several studies with police officers.[53, 193]

In addition to the emotional self-regulation techniques, there are other approaches that also increase HRV coherence. For example, a study of Zen monks found that monks with greater experience in meditation tended to have more coherent heart rhythms during their resting recording than those who had been monks for less than two years.[194] A study of autogenic training showed increased HRV coherence and found that cardiac coherence was strongly correlated with EEG alpha activity. The authors suggested that cardiac coherence could be a general marker for the meditative state.[195] However, this does not suggest that all meditation or prayer styles increase coherence, unless the coherence state is driven by a focus on breathing at a 10-second rhythm or the activation of a positive emotion.[196-199] For example, a study examining HRV while reciting rosary or bead prayers and yoga mantras found that a coherent rhythm was produced by rhythmically breathing, but not by random verbalization or breathing. The authors ascribed the mechanisms for this finding to a breathing pattern of 6-cycles per minute.[200] In a study of the effects of five types of prayer on HRV, it was found that all types of prayer elicited increased cardiac coherence. However, prayers of gratefulness and heartfelt love resulted in definitively higher coherence levels.[201] It also has been shown that tensing the large muscles in the legs in a rhythmical manner at a 10-second rhythm can induce a coherent heart rhythm.[202]

CHAPTER 6

Energetic Communication

The first biomagnetic signal was demonstrated in 1863 by Gerhard Baule and Richard McFee in a magnetocardiogram (MCG) that used magnetic induction coils to detect fields generated by the human heart.[203] A remarkable increase in the sensitivity of biomagnetic measurements has since been achieved with the introduction of the superconducting quantum interference device (SQUID) in the early 1970s. The ECG and MCG signals have since been shown to closely parallel one another.[204]

In this section, we discuss how the magnetic fields produced by the heart are involved in energetic communication, which we also refer to as *cardioelectromagnetic communication*. The heart is the most powerful source of electromagnetic energy in the human body, producing the largest rhythmic electromagnetic field of any of the body's organs. The heart's electrical field is about 60 times greater in amplitude than the electrical activity generated by the brain. This field, measured in the form of an electrocardiogram (ECG), can be detected anywhere on the surface of the body. Furthermore, the magnetic field produced by the heart is more than 100 times greater in strength than the field generated by the brain and can be detected up to 3 feet away from the body, in all directions, using SQUID-based magnetometers (Figure 6.1).

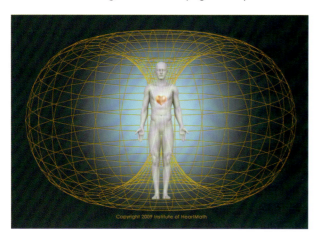

Figure 6.1. The heart's magnetic field, which is the strongest rhythmic field produced by the human body, not only envelops every cell of the body, but also extends out in all directions into the space around us. The heart's magnetic field can be measured several feet away from the body by sensitive magnetometers. Research conducted at HMI suggests the heart's field is an important carrier of information.

Prompted by our findings that the timing between pulses of the heart's magnetic field is modulated by different emotional states, we have performed several studies that show the magnetic signals generated by the heart have the capacity to affect individuals around us.

Biological Encoding of Information

Every cell in our bodies is bathed in an external and internal environment of fluctuating invisible magnetic forces.[205] It has become increasingly apparent that fluctuations in magnetic fields can affect virtually every circuit in biological systems to a greater or lesser degree, depending on the particular biological system and the properties of the magnetic fluctuations.[5, 205] One of the primary ways that signals and messages are encoded and transmitted in physiological systems is in the language of patterns. In the nervous system it is well established that information is encoded in the time intervals between action potentials, or patterns of electrical activity.[206] This also applies to humoral communications in which biologically relevant information also is encoded in the time interval between hormonal pulses.[207-209] As the heart secretes a number of different hormones with each contraction, there is a hormonal pulse pattern that correlates with heart rhythms. In addition to the encoding of information in the space between nerve impulses and in the intervals between hormonal pulses, it is likely that information also is encoded in the interbeat intervals of the pressure and electromagnetic waves produced by the heart. This supports Pribram's proposal discussed earlier that low-frequency oscillations generated by the heart and body in the form of afferent neural, hormonal and electrical patterns are the carriers of emotional information and the higher frequency oscillations found in the EEG reflect the conscious perception and labeling of feelings and emotions.[169] We have proposed that these same rhythmic patterns also can transmit emotional information via the electromagnetic field into the environment, which can be detected by others and processed in the same manner as internally generated signals.

Heartbeat-Evoked Potentials

A useful technique for detecting synchronized activity between systems in biological systems and investigating a number of bioelectromagnetic phenomena is signal averaging. This is accomplished by superimposing any number of equal-length epochs, each of which contains a repeating periodic signal. This emphasizes and distinguishes any signal that is time-locked to the periodic signal while eliminating variations that are not time-locked to the periodic signal. This procedure is commonly used to detect and record cerebral cortical responses to sensory stimulation[210]. When signal averaging is used to detect activity in the EEG that is time-locked to the ECG, the resultant waveform is called the *heartbeat-evoked potential*.

The heart generates a pressure wave that travels rapidly throughout the arteries, much faster than the actual flow of blood that we feel as our pulse. These pressure waves force the blood cells through the capillaries to provide oxygen and nutrients to cells and expand the arteries, causing them to generate a relatively large electrical voltage. These pressure waves also apply pressure to the cells in a rhythmic fashion that can cause some of their proteins to generate an electrical current in response to this "squeeze." Experiments conducted in our laboratory have shown that a change in the brain's electrical activity can be seen when the blood-pressure wave reaches the brain around 240 milliseconds after systole.

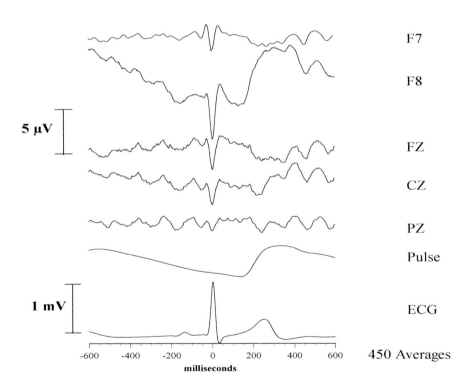

Figure 6.2. Heartbeat-evoked potentials. This figure shows an example of typical heartbeat-evoked potentials. In this example, 450 averages were used. The pulse wave also is shown, indicating the timing relationship of the blood-pressure wave reaching the brain. In this example, there is less synchronized alpha activity immediately after the R-wave. The time range between 10 and 240 milliseconds is when afferent signals from the heart are impinging upon the brain and the alpha desynchronization indicates the processing of this information. Increased alpha activity can be seen later in the waveforms, starting at around the time the blood-pressure wave reaches the brain.

There is a replicable and complex distribution of heartbeat-evoked potentials across the scalp. Changes in these evoked potentials associated with the heart's afferent neurological input to the brain are detectable between 50 to 550 milliseconds after the heartbeat.[8] Gary Schwartz and his colleagues at the University of Arizona believe the earlier components in this complex distribution cannot be explained by simple physiological mechanisms alone and suggest that an energetic interaction between the heart and brain also occurs.[211] They have confirmed our findings that heart-focused attention is associated with increased heart-brain synchrony, providing further support for energetic heart-brain communications.[5] Schwartz and his colleagues also demonstrated that when subjects focused their attention on the perception of their heartbeat, the synchrony in the preventricular region of the heartbeat-evoked potential increased. They concluded that this synchrony may reflect an energetic mechanism of heart-brain communication, while post-ventricular synchrony most likely reflects direct physiological mechanisms.

Biomagnetic Communication Between People

We have found there is a direct relationship between the heart-rhythm patterns and the spectral information encoded in the frequency spectra of the magnetic field radiated by the heart. Thus, information about a person's emotional state is encoded in the heart's magnetic field and is communicated throughout the body and into the external environment.

Figure 6.3 shows two different power spectra derived from an average of 12 individual 10-second epochs of ECG data recorded during differing psychophysiological modes. The plot on the left was produced while the subject was in a state of deep appreciation, whereas the plot on the right was generated while the subject experienced recalled feelings of anger. The difference in the patterns and thus the information they contain, can be seen clearly. There is a direct correlation between the patterns in the heart rate variability rhythm and the frequency patterns in the spectrum of the ECG or MCG. Experiments such as these indicate that psychophysiological information can be encoded into the electromagnetic fields produced by the heart.[163,212]

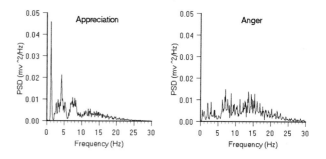

Figure 6.3. **ECG spectra during different emotional states.** The above graphs are the average power spectra of 12 individual 10-second epochs of ECG data, which reflect information patterns contained in the electromagnetic field radiated by the heart. The left-hand graph is an example of a spectrum obtained during a period of high heart-rhythm coherence generated during a sustained heartfelt experience of appreciation. The graph on the right depicts a spectrum associated with a disordered heart rhythm generated during feelings of anger.

The human body is replete with mechanisms for detecting its external environment. Sense organs, the most obvious example, are specifically geared to react to touch, temperature, select ranges of light, sound waves, etc. These organs are acutely sensitive to external stimuli. The nose, for example, can detect one molecule of gas, while a cell in the retina of the eye can detect a single photon of light. If the ear were any more sensitive, it would pick up the sound of the random vibrations of its own molecules.[213]

The interaction between two human beings, such as, the consultation between patient and clinician or a discussion between friends, is a very sophisticated dance that involves many subtle factors. Most people tend to think of communication solely in terms of overt signals expressed through facial movements, voice qualities, gestures and body movements. However, evidence now supports the perspective that a subtle yet influential electromagnetic or "energetic" communication system operates just below our conscious level of awareness. The following section will discuss data that suggests this energetic system contributes to the "magnetic" attractions or repulsions that occur between individuals.

The ability to sense what other people are feeling is an important factor in allowing us to connect, or communicate effectively with them. The smoothness or flow in any social interaction depends to a great extent on the establishment of a spontaneous entrainment or linkage between individuals. When people are engaged in deep conversation, they begin to fall into a subtle dance, synchronizing their movements and postures, vocal pitch, speaking rates and length of pauses between responses,[214] and, as we are now discovering, important aspects of their physiology also can become linked and synchronized.

The Electricity of Touch: Detection and Measurement of Cardiac Energy Exchange Between People

An important step in testing our hypothesis that the heart's electromagnetic field could transmit signals between people was to determine if an individual's field and the information modulated within it could be detected by others. In conducting these experiments, the question being asked was straightforward: Can the electromagnetic field generated by the heart of one individual be detected in physiologically relevant ways in another person, and if so, does it have any discernible biological effects? To investigate these possibilities, we used signal-averaging techniques to detect signals that were synchronous with the peak of the R-wave of one subject's ECG in recordings of another subject's electroencephalogram (EEG) or brain waves. My colleagues and I have performed numerous experiments in our laboratory over several years using these techniques.[215] Several examples are included below to illustrate some of our findings. In the majority of these experiments, subjects were seated in comfortable, high-back chairs to minimize postural changes with the positive ECG electrode located on the side at the left sixth rib and referenced to the right supraclavicular fossa, according to the International 10-20 system. The ECG and EEG were recorded for both subjects simultaneously so the data (typically sampled at 256 hertz or higher) could be analyzed for simultaneous signal detection in both (Figure 6.4).

To clarify the direction in which the signal flow was analyzed, the subject whose ECG R-wave was used as the time reference for the signal-averaging procedure is referred to as the "signal source," or simply "source." The subject whose EEG was analyzed for the registration of the source's ECG signal is referred to as the "signal receiver," or simply "receiver." The number of averages used in the majority of the experiments was 250 ECG cycles (~4 minutes). The subjects did not consciously intend to send or receive a signal and, in most cases, were unaware of the true purpose of the experiments. The results of these experiments have led us to conclude that the nervous system acts as an antenna, which is tuned to and responds to the magnetic fields produced by the hearts of other individuals. My colleagues and I call this energetic information exchange *energetic communication* and believe it to be an innate ability that heightens awareness and mediates important aspects of true empathy and sensitivity to others. Furthermore, we have observed that this energetic communication ability can be enhanced, resulting in a much deeper level of nonverbal communication, understanding and connection between people. We also propose that this type of energetic communication between individuals may play a role in therapeutic interactions between clinicians and patients that has the potential to promote the healing process.

From an electrophysiological perspective, it appears that sensitivity to this form of energetic communication between individuals is related to the ability to be emotionally and physiologically coherent. The data indicate that when individuals are in the coherent state, they are more sensitive to receiving information contained in the magnetic fields generated by others. In addition, during physiological coherence, internal systems are more stable, function more efficiently and radiate electromagnetic fields containing a more coherent structure.[163]

The first step was to determine if the ECG signal of one person could be detected in another individual's EEG during physical contact. For these experiments, we seated pairs of subjects 4 feet apart and monitored them simultaneously.

Although in most pairs a clear signal transfer between the two subjects was measurable in one direction, it was only observed in both directions simultaneously in about 30 percent of the pairs (i.e., Subject 2's ECG could be detected in Subject 1's EEG at the same time Subject 1's ECG was detectable in Subject 2's EEG). As shown later, an important variable appears to be the degree of physiological coherence maintained. After demonstrating that the heart's activity could be detected in another's EEG during physical contact, we completed a series of experiments to determine if the signal was transferred via electrical conduction alone or if it also was energetically transferred via magnetic fields. The results suggest a significant degree of signal transfer occurs through skin conduction, but it also is radiated between individuals, which will be discussed next.

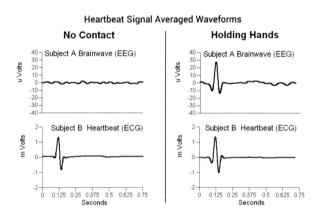

Figure 6.4. Heartbeat signal-averaged waveforms showing a transference of the electrical energy generated by Subject B's heart can be detected in Subject A's EEG (brain waves) when they hold hands.

Heart-Brain Synchronization During Nonphysical Contact

Because the magnetic component of the field produced by the heartbeat naturally radiates outside the body and can be detected several feet away with SQUID-based magnetometers,[217] we decided to further test the transference of signals between subjects who were not in physical contact. In these experiments, the subjects either were seated side by side or facing each other at varying distances. In some cases, we were able to detect a clear QRS-shaped signal in the receiver's EEG. Although the ability to obtain a clear registration of the ECG in the other person's EEG declined as the distance between subjects was increased, the phenomenon appears to be nonlinear. For instance, a clear signal could be detected at a distance of 18 inches in one session, but was undetectable in the very next trial at a distance of only 6 inches. Although transmission of a clear QRS-shaped signal is uncommon at distances over 6 inches in our experience, physiologically relevant information is communicated between people at much further distances and is reflected in synchronized activity.

Figure 6.5 shows the data from two subjects seated and facing one another at a distance of 5 feet, with no physical contact. They were asked to use the Heart Lock-In Technique,[179] which has been shown to produce sustained states of physiological coherence.[116] Participants were not aware of the purpose of the experiment. The top three traces show the signal-averaged waveforms derived from the EEG locations along the medial line of the head.

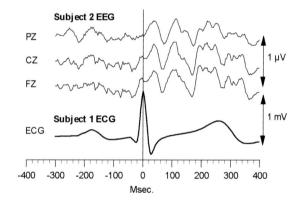

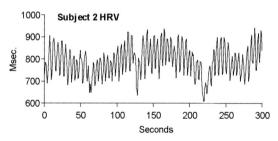

Figure 6.5. Heart-brain synchronization between two people. The top three traces are Subject 2's signal-averaged EEG waveforms, which are synchronized to the R-wave of Subject 1's ECG. The lower plot shows Subject 2's heart rate variability pattern, which was coherent throughout the majority of the record. The two subjects were seated at a conversational distance without physical contact.

Note that in this example the signal-averaged waveforms do not contain any semblance of the QRS complex shape as seen in the physical contact experiments. Rather, they reveal the occurrence of an alpha-wave synchronization in the EEG of one subject that is precisely timed to the R-wave of the other subject's ECG.

Power-spectrum analysis of the signal-averaged EEG waveforms showed that the alpha rhythm was synchronized to the other person's heart. This alpha synchronization does not imply that there is increased alpha activity, but it does show that the existing alpha rhythm is able to synchronize to extremely weak external electromagnetic fields such as those produced by another person's heart. It is well known that the alpha rhythm can synchronize to an external stimulus such as sound or light flashes, but the ability to synchronize to such a subtle electromagnetic signal is surprising. As mentioned, there also is a significant ratio of alpha activity that is synchronized to one's own heartbeat and the amount of this synchronized alpha activity is significantly increased during periods of physiological coherence.[5, 219]

Figure 6.6 shows an overlay plot of one of Subject 2's signal-averaged EEG traces and Subject 1's signal-averaged ECG.

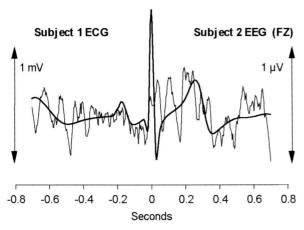

Figure 6.6. Overlay of signal-averaged EEG and ECG. This graph is an overlay plot of the same EEG and ECG data shown in Figure 6.5. Note the similarity of the wave shapes, indicating a high degree of synchronization.

This view shows an amazing degree of synchronization between the EEG of Subject 2 and Subject 1's heart. These data show it is possible for the magnetic signals radiated by the heart of one individual to influence the brain rhythms of another. In addition, this phenomenon can occur at conversational distances.

Energetic Sensitivity and Empathy

Figure 6.7 shows the data from the same two subjects during the same time period, but it is analyzed for alpha synchronization in the opposite direction (Subject 1's EEG and Subject 2's ECG). In this case, we see that there is no observable synchronization between Subject 1's EEG and Subject 2's ECG. The key difference between the data shown in figures 6.5 and 6.6 is the high degree of physiological coherence maintained by Subject 2. In other words, the degree of coherence in the receiver's heart rhythms appears to determine whether his/her brain waves synchronize to the other person's heart.

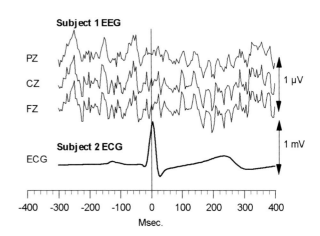

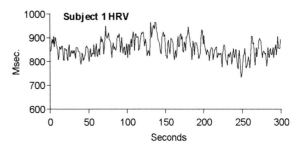

Figure 6.7. The top three traces are the signal-averaged EEG waveforms for Subject 1. There is no apparent synchronization of Subject 1's alpha rhythm to Subject 2's ECG. The bottom plot is a sample of Subject 1's heart rate variability pattern, which was incoherent throughout the majority of the record.

This suggests that when a person is in a physiologically coherent state, he or she exhibits greater sensitivity in registering the electromagnetic signals and information patterns encoded in the fields radiated by others' hearts. At first glance the data may be interpreted to mean we are more vulnerable to the potential negative influence of incoherent patterns radiated by those around us. In fact, the opposite is true. When people are able to maintain the physiological coherence mode, they are more internally stable and thus less vulnerable to being negatively affected by the fields emanating from others. It appears that increased internal stability and coherence is what allows the increased sensitivity to emerge.

This fits quite well with our experience in training thousands of individuals how to self-generate and maintain coherence while they are communicating with others. Once individuals learn this skill, it is a common experience that they become much more attuned to other people and are able to detect and understand the deeper meaning behind spoken words. They often are able to sense what someone else truly wishes to communicate even when the other person may not be clear in what he or she is attempting to say. The Coherent Communication Technique helps people to feel fully heard, speak authentically and with discernment and promote greater rapport and empathy between people.[180]

Heart-Rhythm Synchronization Between People

When heart rhythms are more coherent, the electromagnetic field that is radiated outside the body correspondingly becomes more organized, as shown in Figure 6.3. The data presented thus far indicate that signals and information can be communicated energetically between individuals and that they have measurable biological effects, but so far have not implied a literal synchronization of two individuals' heart-rhythm patterns. We have found that synchronization of heart-rhythm patterns between individuals is possible, but usually occurs only under specific conditions. In our experience, true heart-rhythm synchronization between individuals is rare during normal waking states. We have found that individuals who have a close working or living relationship are the best candidates for exhibiting true heart-rhythm synchronization. Figure 6.8 shows an example of heart-rhythm synchronization between two women who have a close working relationship and practice coherence-building techniques regularly. For this experiment, they were seated 4 feet apart and were consciously focused on generating feelings of appreciation for each other.

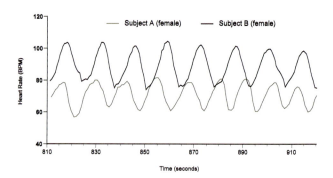

Figure 6.8. Heart-rhythm entrainment between two people. These data were recorded while both subjects were practicing the Heart Lock-In Technique and consciously feeling appreciation for each other.

A more complex type of synchronization also can occur during sleep. Although we have only looked at couples who are in long-term stable and loving relationships, we have been surprised at the high degree of heart-rhythm synchrony observed in these couples while they sleep. Figure 6.9 shows an example of a small segment of data from one couple.

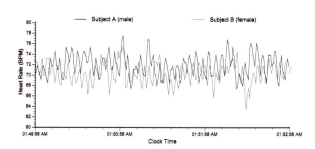

Figure 6.9. Heart-rhythm entrainment between husband and wife during sleep.

These data were recorded using an ambulatory ECG recorder with a modified cable harness that allowed the concurrent recording of two individuals on the same recording. Note how the heart rhythms simultaneously change in the same direction and how heart rates converge. Throughout the recording, clear transition periods are evident in which the heart rhythms move into greater synchronicity for some time and then drift out again. This implies that unlike in most wakeful states, synchronization between the heart rhythms of individuals can and does occur during sleep.

Another line of research that has shown physiological synchronization between people was in a study of a 30-minute Spanish firewalking ritual. Heart-rate data was obtained from 38 participants and synchronized activity was compared between firewalkers and spectators. They showed fine-grained commonalities of arousal during the ritual between firewalkers and related spectators but not unrelated spectators. The authors concluded that their findings demonstrated that a collective ritual can evoke synchronized arousal over time between active participants and relatives or close friends. They also suggest that the study links field observations to a physiological basis and offers a unique approach for the quantification of social effects on human physiology during real-world interactions, a mediating mechanism that is likely informational.[220]

Morris[221] studied the effect of heart coherence in a group setting with participants who were trained in HeartMath's Quick Coherence® Technique. He conducted 148 10-minute trials in which three trained participants were seated around a table with one untrained participant. During each trial, three of the trained participants were placed with untrained volunteers to determine whether the three could collectively facilitate higher levels of HRV coherence in the untrained individual. The coherence of the HRV of the untrained subject was found to be higher in approximately half of all matched comparisons when the trained participants focused on achieving increased coherence. In addition, evidence of heart-rhythm synchronization between group participants was revealed through several evaluation methods and higher levels of coherence correlated to higher levels of synchronization between participants. There was a statistical relationship between this synchronization and relational measures (bonding) among the participants. The authors concluded that "evidence of heart-to-heart synchronization across subjects was found, lending credence to the possibility of heart-to-heart biocommunications."

Using signal-averaging techniques, we also were able to detect synchronization between a mother's brain waves (EEG-CZ) and her baby's heartbeats (ECG). The pair were not in physical contact, but when the mother focused her attention on the baby, her brain waves synchronized to the baby's heartbeats (Figure 6.10). We were not able to detect that the infant's EEG synchronized to the mother's heartbeats.

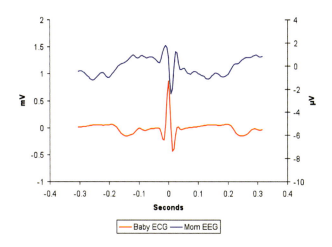

Figure 6.10. ECG and EEG synchronization between mother and baby.

Biomagnetic Communication Between People and Animals

Farmers and attentive observers know that most cattle and sheep, when grazing, face the same way. It has been demonstrated by means of satellite images, field observations and measurements of deer beds in snow that domestic cattle across the globe and grazing and resting red and roe deer align their body axes in roughly a north-south direction and orient their heads northward when grazing or resting. Wind and light conditions were excluded as common determining factors, so magnetic alignment with the earth's geomagnetic field was determined to be the best explanation. Magnetic north was a better predictor than geographic north, suggesting large mammals have magnetoreception capability.[222]

Science of the Heart

We also have found that a type of heart-rhythm synchronization can occur in interactions between people and their pets. Figure 6.11 shows the results of an experiment looking at the heart rhythms of my son, Josh (age 12 at the time of the recording) and his dog, Mabel. Here we used two Holter recorders, one fitted on Mabel and the other on Josh. We synchronized the recorders and placed Mabel in one of our labs.

Josh entered the room and sat down and proceeded to do a Heart Lock-In and consciously radiate feelings of love toward Mabel. There was no physical contact and he did not make any attempts to get the dog's attention. In Figure 6.11, note the synchronous shift to increased coherence in the heart rhythms of both Josh and Mabel as Josh consciously feels love for his pet.

Another example of an animal's heart-rhythm pattern shifting in response to a human's shift of emotional states is shown in Figure 6.12. This was a collaborative study with Ellen Gehrke, Ph.D. who consciously shifted into a coherent state while sitting in a corral with her horse, neither touching nor petting it. When she shifted into a coherent state, the horse's heart-rhythm pattern also shifted to a more ordered pattern.

In other trials, very similar shifts in horses' HRV patterns were seen in three out of four horses' heart rhythms. One of the horses that did not show any response was well known for not relating well to humans or other horses.

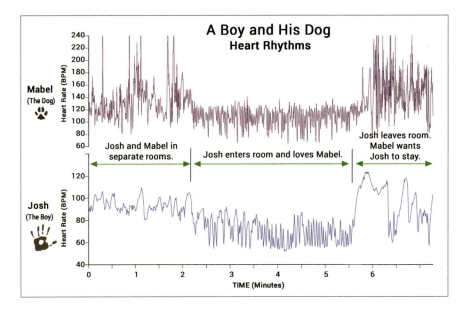

Figure 6.11. Heart-rhythm patterns of a boy and his dog. These data were obtained using ambulatory ECG recorders fitted on both Josh, a young boy and Mabel, his pet dog. When Josh entered the room where Mabel was waiting and consciously felt feelings of love and care towards his pet, his heart rhythms became more coherent and this change appears to have influenced Mabel heart rhythms, which shifted to a more coherent rhythm.

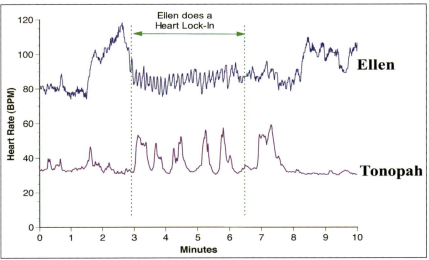

Figure 6.12. Heart-rhythm patterns of woman and horse. These data were obtained using ambulatory ECG recorders fitted on both Ellen and her horse, Tonopah. When she did a Heart Lock-In, her heart rhythms became more coherent and this change appears to have influenced the horse's heart rhythms.

© Copyright 2015 HeartMath Institute

CHAPTER 7

Intuition Research: Coherence and the Surprising Role of the Heart

In a world whose pace and complexity is ever increasing, where we face more and more personal and social challenges, there is a call for lifting individual and global consciousness. People want to make more intelligent choices so daily life is more tenable, personal and global relationships are stronger and more meaningful and the future of our planet is assured.[223]

Raising individual and global consciousness can help us improve personal and collective health, well-being and harmony. We suggest that this begins with people taking greater responsibility for their day-to-day decisions, actions and behaviors, which can result in establishing a new and healthier physiological and psychological internal baseline reference. Establishing a baseline requires having effective and practical strategies available for handling daily situations, making good decisions and taking meaningful and appropriate action.

Much attention has been given to identifying the many factors that go into making good decisions. Among these are awareness of self and others, cognitive flexibility and self-regulation of emotions. All of these are important for bringing more consciousness into our daily situations and the decisions we make. Something else that should be considered in good decision-making – and we've all experienced it, perhaps without being fully aware of it – is intuition. There is fascinating research that is beginning to uncover the nature and function of intuition, or what researchers refer to as intuitive intelligence. In a literature review of intuition, Gerard Hodgkinson of Leeds University in England notes that despite the many conceptualizations of intuition, there is a growing body of research suggesting there are underlying nonconscious aspects of intuition. Among the nonconscious aspects of intuition which are involved in intuitive perception are implicit learning, or implicit knowledge.[224] It is commonly acknowledged that intuitive perception plays an important role in business decisions and entrepreneurship, learning, medical diagnosis, healing, spiritual growth and overall well-being.[225, 226]

Research also suggests intuition may play an important role in social cognition, decision-making and creativity. When addressing life situations, people often default to familiar patterns of thoughts, feelings and actions in both the decision-making process and how they see others.

Rather than responding to situations from habitual patterns that are not necessarily healthy or constructive, those situations could be more effectively addressed with new and creative solutions. These solutions can take into consideration the available inner resources that are congruent with one's deeper intuition and core values. In other words, we can learn to intentionally align with and access our intuitive intelligence, which can provide moment-to-moment guidance and empower what HeartMath calls heart-based living, reliance in all things on the wisdom, intelligence and qualities of the heart.

The origin of the word "intuition" is the Latin verb *intueri*, which is usually translated as to look inside or to contemplate. Hodgkinson concludes that "intuiting" is a complex set of interrelated cognitive, affective and somatic processes in which there is no apparent intrusion of deliberate, rational thought. He also concludes that the considerable body of theory and research that has emerged in recent years demonstrates that the concept of intuition has emerged as a legitimate subject of scientific inquiry that has important ramifications for educational, personal, medical and organizational decision-making, personnel selection and assessment, team dynamics, training and organizational development.[224] Another comprehensive review of intuition literature yielded this definition of intuition: "Affectively charged judgments that arise through rapid, nonconscious and holistic associations."[227]

Several researchers have contended that intuition is an innate ability that all humans possess in one form or another and is arguably the most universal natural ability we possess. They also say the ability to intuit could be regarded as an inherited unlearned gift.[228, 229] A common element also found in most discussions and definitions of intuition is that of affect or emotions. Although intuitions are felt, they can be accompanied by cognitive content and perception of information. Our research and experience suggests that emotions are the primary language of intuition and that intuition offers a largely untapped resource to manage and uplift our emotions, daily experience and consciousness.

Types of Intuition

Our research at the HeartMath Institute suggests there are three categories or types of processes the word intuition describes. The first type of intuition, often called *implicit knowledge* or *implicit learning*, essentially refers to knowledge we've acquired in the past and either forgot or did not realize we had learned. Drawing on the neuroscience conception of the human brain as a highly efficient and effective pattern-matching device,[176] a number of pattern-recognition models have been developed to show how this fast type of intuitive decision-making and action can be understood purely in terms of neural processes. In this regard, the brain matches the patterns of new problems or challenges with implicit memories based on prior experience.[224, 230, 231]

The second type of intuition is what we call *energetic sensitivity*, which refers to the ability of the nervous system to detect and respond to environmental signals such as electromagnetic fields (also see Energetic Communication section). It is well established that in both humans and animals, nervous-system activity is affected by geomagnetic activity.[232] Some people, for example, appear to have the capacity to feel or sense that an earthquake is about to occur before it happens. It has recently been shown that changes in the earth's magnetic field can be detected about an hour or even longer before a large earthquake occurs.[231] Another example of energetic sensitivity is the sense of being stared at. Several scientific studies have verified this type of sensitivity.[234]

The third type of intuition is *nonlocal intuition*, which refers to the knowledge or sense of something that cannot be explained by past or forgotten knowledge or by sensing environmental signals. It has been suggested that the capacity to receive and process information about nonlocal events appears to be a property of all physical and biological organization and this likely is because of an inherent interconnectedness of everything in the universe.[235-237] Examples of nonlocal intuition include when a parent senses something is happening to his or her child who is many miles away, or the repeated, successful sensing experienced by entrepreneurs about factors related to making effective business decisions.

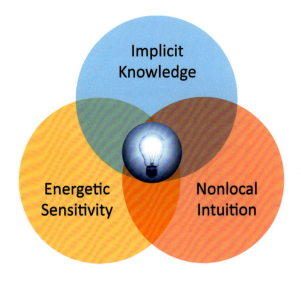

Figure 7.1. The three types of intuition

Implicit Learning

The question of how intuition interacts with deliberate, conscious thought processes, has long been the subject of debate. Research in the fields of cognitive and social psychology has produced the commonly accepted dual-process theory, which suggests there are two separate processing systems. The first system is unconscious, automatic and intuitive. It processes information very rapidly and associates current inputs to the brain with past experiences. Therefore, it is relatively undemanding in its use of cognitive resources. For example, when individuals have gained experience in a particular field, implicit intuitions are derived from their capacity to recognize important environmental cues and rapidly and unconsciously match those cues to existing familiar patterns. This results in rapid and effective diagnosis or problem-solving. In contrast, the second processing system is conscious in nature, relatively slow, rule-based and analytic. It places greater demands on cognitive resources than the first system.[224]

Insight

The term intuition also is used commonly to describe experiences scientific literature refers to as *insight*. When we have a problem we cannot immediately solve, the brain can be working on it subconsciously. It is common when we are in the shower, driving or doing something else and not thinking about the problem that a solution pops into the conscious mind, a process we experience as an *intuitive insight*. This type of implicit process involves a longer gestation period following an impasse in problem-solving before a sudden insightful perception or strategy that leads to a solution.[238] In contrast, intuition in the dual-processing models of implicit intuition described above occurs almost instantaneously and is emotionally charged.[239]

Nonlocal Intuition

The study of nonlocal intuition, which at times has been thought of as being in the same category as telepathy, clairvoyance and precognition, has been fraught with debate in the scientific community.[240] While there are various theories that attempt to explain how the process of intuition functions, these theories have yet to be confirmed, so an integrated theory remains to be formulated. Nevertheless, there is now a large body of documented rigorous scientific research on nonlocal intuitive perception that dates back more than seven decades. A variety of experiments show it cannot be explained by flaws in experimental design or research methods, statistical techniques, chance or selective reporting of results.[239]

A meta-analysis of nine experiments that measured physiological responses occurring before a future event (pre-stimulus responses) that could not otherwise be anticipated through any known inferential process, revealed statistically significant results in eight of the nine studies in over 1,000 subjects.[240] Subsequent to this, a researcher, by examining 26 studies, also concluded that a clear pre-stimulus response in physiological activity occurred before unpredictable stimuli, despite the fact there is not yet any known explanation of the mechanisms for this finding.[242]

There is compelling evidence to suggest the physical heart is coupled to a field of information not bound by the classical limits of time and space.[243, 244] This evidence comes from a rigorous experimental study that demonstrated the heart receives and processes information about a future event before the event actually happens.[243, 244]

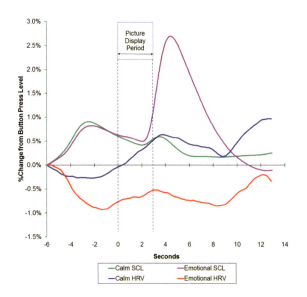

Figure 7.2. The heart's pre-stimulus response. The graph shows group averages of the heart rate variability (blue and red lines) and skin conductance level (pink and green lines) responses. The "0" time point denotes when the photos were first shown, when participants saw either an emotionally arousing or calm picture. Pre-stimulus responses which indicate nonlocal intuition are in the period between −6 and 0 seconds. The red line is the HRV trace when the future photo was an emotional one, and the blue line shows the HRV for calming future photos. The highly significant difference between the HRV responses in the pre-stimulus period before the future calm or emotional photos can clearly be seen starting to diverge approximately 4.8 seconds prior to the participants actually seeing the photos.

Extending and building on Radin's protocol designed to evoke an emotional response using randomly selected, emotionally arousing or calming photographs, we added measures of brain response (EEG) and heart-rhythm activity (ECG) and found that not only did both the brain and heart receive the pre-stimulus information some 4 to 5 seconds before a future emotional picture was randomly selected by the computer, the heart actually received this information about 1.5 seconds before the brain received it (Figure 7.3).[244]

A number of studies have since found evidence of the heart's role in reflecting future or distant events.[245-251] Using a combination of cortical-evoked potentials and heartbeat-evoked potentials, these studies also found that when the participants were in the physiological coherence mode before the trials, the afferent input from the heart and cardiovascular system modulated changes in the brain's electrical activity, especially at the frontal areas of the brain. In other words, participants were more attuned to information from the heart while in a coherent state before participating in the experimental protocol. Therefore, being in a state of psychophysiological coherence is expected to enhance intuitive ability.[244]

This suggests the heart is directly coupled to a source of information that interacts with the multiplicity of energetic fields in which the body is embedded. We also found further evidence that the magnitude of pre-stimulus response to a future event is related to the degree of emotionality associated with that event.[243]

Nonlocal Intuition in Repeat Entrepreneurs

A study conducted in Iran with a group of 30 repeat entrepreneurs in the science and technology parks of the city of Tehran duplicated and extended our first study of intuition.[251] Repeat entrepreneurs were chosen for this study because they are most likely to have demonstrated that their success is not the result of luck alone and they have beaten the odds against success. Also, studies have shown that they have a strong tendency to rely on their intuitions when making important business decisions. The study was modeled after our study, described above, whose stimulus was a computer-administered random sequence of calm and emotional pictures. However, this study added a new element: Researchers conducted two separate experiments.

The first, with a group of single participants (N = 15), and the second, with a group of co-participant pairs (N = 30), investigated the "amplification" of intuition effects by social connection. In the experiment for single participants, the participant watched the pictures on a monitor alone, while in the experiment for

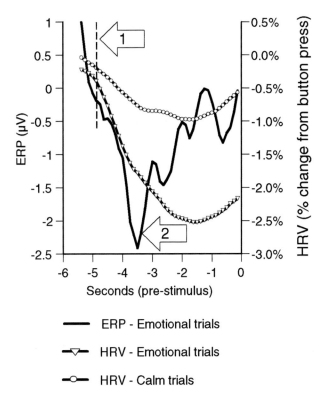

Figure 7.3. **Example of temporal dynamics of heart and brain pre-stimulus responses:** This overlay plot shows the mean event-related potential (ERP) at EEG site FP2 and heart-rate deceleration curves during the pre-stimulus period. (The "0" time point denotes stimulus onset.) The heart-rate deceleration curve for the trials, in which a negative emotionally arousing photo would be seen in the future, diverged from that of trials containing a calming future picture (sharp downward shift) about 4.8 seconds before the stimulus (arrow 1). The emotional trials ERP showed a sharp positive shift about 3.5 seconds before the stimulus (arrow 2). This positive shift in the ERP indicates when the brain "knew" the nature of the future stimulus. The time difference between these two events suggests that the heart received the intuitive information about 1.3 seconds before the brain. Heartbeat-evoked potential analysis confirmed that a different afferent signal was sent by the heart to the brain during this period.[244]

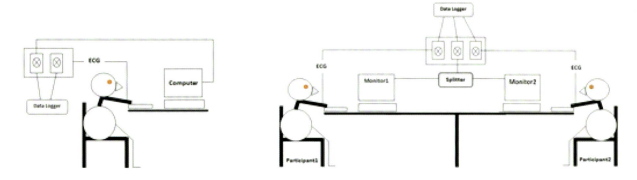

Figure 7.4. Setup for Single Participant Experiment and Co-participant Pair Experiment.

co-participant pairs, each pair watched the same pictures simultaneously on two monitors while sitting facing each other, as illustrated in Figure 7.4.

Each experiment was conducted over 45 trials while heart-rate rhythm activity was recorded continuously. In both experiments, the results showed significant pre-stimulus results, meaning for the period before the computer had randomly selected the picture stimulus. Moreover, while significant separation between the emotional and calm HRV curves was observed in the single-participant experiment, an even larger separation was apparent for the experiment with co-participant pairs, and the difference between the two groups also was significant. Overall, the results of the single-participant experiment confirm our and others' previous finding that electrophysiological measures, especially changes in heart rhythm, can demonstrate intuitive foreknowledge. This result is notable because, having come from experiments in Iran, it constituted cross-cultural corroboration in a non-Western context. In addition, the results for co-participant pairs offer new evidence on the amplification of the nonlocal intuition signal.

Full-Moon Effect on Amplifying Intuition

We also evaluated an updated version of a roulette protocol we developed that includes two pre-stimulus segments. This study included an analysis of individual data analysis and group-level data analysis for 13 participants over eight separate trials.[252] We also assessed the potential effects of the moon phase on the pre-stimulus response outcomes and participant winning and amount-won ratios. Half of the experimental sessions were conducted during the full-moon phase and half during the new-moon phase. Within each trial, a total of three segments of physiological data were assessed. There were two separate pre-stimulus periods, a pre-bet (4-seconds) and post-bet (12-seconds), and a post-result period (6-seconds). Participants were told they were participating in a gambling experiment, given an initial starting kitty and informed that they could keep any winnings over the course of 26 trials for each of the eight sessions. The physiological measures included the ECG, from which cardiac interbeat intervals (HRV) were derived and skin conductance.

Overall, the results indicate the protocol provides an effective objective method for measuring and detecting a pre-stimulus response, which demonstrates a type of nonlocal intuition. We found significant differences between the win and loss responses in the aggregated physiological waveform data during both pre-stimulus segments (Figure 7.5).

On average, we detected a significant pre-stimulus response starting around 18 seconds before participants knew the future outcome. Interestingly, there was not a strong overall relationship between the pre-stimulus responses and the amount of money the participants won or lost. We also found a significant difference in both pre-stimulus periods during the full-moon phase, when they also won more money, but not in the new-moon phase (Figure 7.6). Overall, the findings also suggest that if participants had been able to become more attuned to their internal cardiac related pre-stimulus responses, they would have performed much better on the betting choices they made.

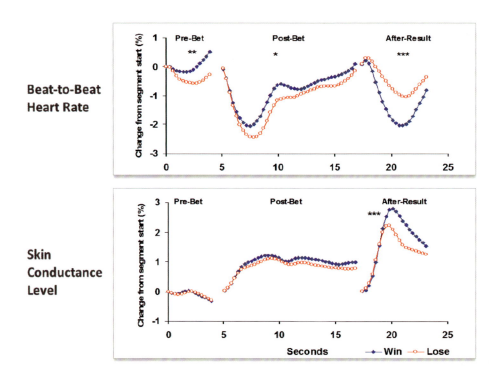

Figure 7.5. Multisession roulette paradigm study: Grand averages are shown for the skin conductance levels and HRV win/loss waveform differences in response to winning or losing for all 13 participants across all eight trials for the three segments of the experiment: pre-bet, post-bet and post-result periods. * = (p < 0.05), ** = (p < 0.01), *** = (p < 0.001)

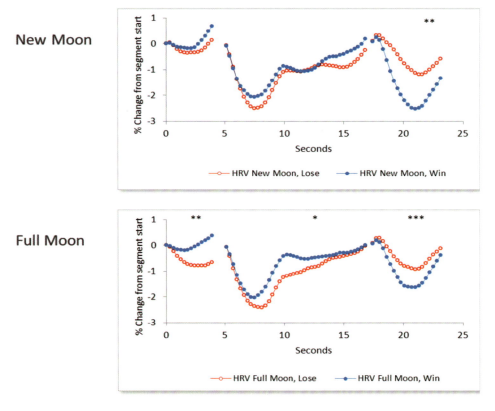

Figure 7.6. Multisession roulette paradigm study: Shown are grand averages by moon phase for the skin conductance levels and HRV win/loss waveforms results. There was a significant difference in the HRV win/loss waveforms in both the pre-bet (p < 0.01) and post-bet (p < 0.05) periods during the full-moon phase and no significant difference during the new-moon phase. * = (p < 0.05), ** = (p < 0.01), *** = (p < 0.001)

Heart Intelligence

Because the heart plays a central role in creating physiological coherence and is associated with heartfelt positive emotions and intuition, it is not surprising that one of the strongest threads uniting the views of diverse cultures and religious and spiritual traditions throughout history has been a universal regard that it is the source of love, wisdom, intuition, courage, etc. Everyone is familiar with such expressions as "put your heart into it," "learn it by heart" and "speak from your heart." All of these suggest an implicit knowledge that the heart is more than a physical pump that sustains life. Such expressions reflect what often is called the intuitive, or spiritual heart. Throughout history, people have turned to the intuitive heart – also referred to as their inner voice, soul or higher power – as a source of wisdom and guidance.

We suggest that the terms intuitive heart and spiritual heart refer to our *energetic heart*, which we believe is coupled with a deeper part of ourselves. Many refer to this as their higher self or higher capacities, or what physicist David Bohm described as "our implicate order and undivided wholeness."[235] We use the term *energetic systems* in this context to refer to the functions we cannot directly measure, touch or see, such as our emotions, thoughts and intuitions. Although these functions have loose correlations with biological activity patterns, they nevertheless remain covert and hidden from direct observation. Several notable scientists have proposed that such functions operate primarily in the frequency domain outside of time and space and they have suggested some of the possible mechanisms that govern how they are able to interact with biological processes.[206, 253-259]

As discussed in the Heart-Brain Communication chapter of this work, the physical heart has extensive afferent connections to the brain and can modulate perception and emotional experience.[5] Our experience suggests that the physical heart also has communication channels connecting it with the energetic heart.[244] Nonlocal intuition, therefore, is transformational, and from our perspective, it contains the wisdom that streams from the soul's higher information field down into the psychophysiological system via the energetic heart and can inform our moment-to-moment experiences and interactions. At HeartMath Institute, this is what we call *heart intelligence*.

Heart intelligence is the flow of higher awareness and the intuition we experience when the mind and emotions are brought into synchronistic alignment with the energetic heart. When we are heart-centered and coherent, we have a tighter coupling and closer alignment with our deeper source of intuitive intelligence. We are able to more intelligently self-regulate our thoughts and emotions and over time this lifts consciousness and establishes a new internal physiological and psychological baseline.[244] In other words, there is an increased flow of intuitive information that is communicated via the emotional energetic system to the mind and brain systems, resulting in a stronger connection with our deeper inner voice.

Accessing Intuition

Although people's degree of access to the heart's intuition varies, we all have access to the three types of intuition. As we learn to slow down our minds and attune to our deeper heart feelings, a natural intuitive connection can occur. Intuition often is thought of in the context of inventing a new lightbulb or winning in Las Vegas, but what most people discover is that intuition is a very practical asset that can help guide their moment-to-moment choices and decisions in daily life. Our intuitive insights often unfold more understanding of ourselves, others, issues and life than years of accumulated knowledge. It is especially helpful for eliminating unnecessary energy expenditures, which deplete our internal reserves, making it more difficult to self-regulate and be in charge of our attitudes, emotions and behaviors in ordinary day-to-day life situations. Intuition allows us to increase our ability to move beyond automatic reactions and perceptions. It helps us make more intelligent decisions from a deeper source of wisdom, intelligence and balanced discernment, in essence increasing our consciousness, happiness and the quality of our life experience. This increases synchronicities and enhances our creativity and ability to flow through life. It also increases our ability to handle awkward situations such as dealing with difficult people with more ease and it promotes harmonious interaction and connectivity with others.

It is important to understand that conscious awareness of anything, including our emotions and intuitive promptings, is not possible until something has captured our attention.[260] Sensory neurons in our eyes, ears, nose and body are continuously active day and night, whether we are awake or asleep. The brain receives a steady stream of information about all the events the sensory systems are detecting. It would be bewildering if we were continuously aware of all the incoming information from both our external and internal environments. In fact, we completely ignore most of the information arriving to the brain – most of the time. It is when inputs are large, sudden or novel or lead to an emotional reaction that they capture and focus our attention and that we become aware of them.[206]

Voluntary attention, on the other hand, describes the process in which we can consciously self-regulate and determine the contents of our own awareness as well as the duration of our focus. Current evidence suggests that this self-regulatory capacity relies on an inner resource akin to energy, which is used to interrupt the stream of consciousness and behavior and alter it. When this limited energy has been depleted, further efforts at self-regulation are less successful than usual.[261] With practice, however, the capacity to self-regulate can be increased and give us more energy resources to sustain self-directed control. Importantly, these practices also are keys to establishing a new baseline and once a new baseline is established, the new patterns of self-regulation become automatic and therefore do not require the same energy expenditure.

One of the most important keys to accessing more of our intuitive intelligence and inner sense of knowing is developing deeper levels of self-awareness of our more subtle feelings and perceptions, which otherwise never rise to conscious awareness. In other words, we have to pay attention to the intuitive signals that often are under the radar of conscious perception or are drowned out by ongoing mental chatter and emotional unrest. A common report from people who practice being more self-aware of their inner signals is that the heart communicates a steady stream of intuitive information to the mind and brain. In many cases, we only perceive a small percentage of intuitive information or choose to override the signals because they do not match our more egocentric desires.

Given that there is a relationship between increased heart coherence and access to intuitive signals,[244] the capacity to shift into a coherent state is an important factor in the three types of intuition: implicit knowledge/learning, energetic sensitivity and nonlocal intuition. The research discussed above suggests it's possible to access intuitive intelligence more effectively by first getting into a coherent state, quieting mental chatter and emotional unrest and paying attention to shifts in our feelings, a process that brings intuitive signals to conscious awareness.[262] We have found that increased heart-rhythm coherence correlates with significant improvements in performance on tasks requiring attentional focus and subtle discrimination.[5] We've also found that heart-rhythm coherence correlates with pre-stimulus-related afferent (ascending) signals from the heart to the brain.[244]

It is likely that these signals are important elements of intuition that are particularly salient in pattern recognition and that they are involved in all types of intuitive processes.

The Freeze Frame Technique,[179, 182] is a five-step process that was designed for improving intuitive capacities, stopping energy drains, shifting perspective, obtaining greater clarity and finding innovative solutions to problems or issues.

CHAPTER 8

Health Outcome Studies

"Natural forces within us are the true healers of disease." —Hippocrates

An estimated 60% to 80% of primary-care doctor visits are related to stress.[60-62] HeartMath's easily learned mental and emotion self-regulation techniques and practices can provide an effective strategy for stress reduction in many clinical contexts. As discussed earlier, these intentionally simple techniques allow people to quickly self-induce a physiological shift to a more coherent state that takes advantage of the concurrent change in afferent neuronal input to the brain, which is associated with increased self-regulatory capacity and thus, ability to more successfully handle the demands and challenges of life with more ease and composure. Consequently, there is a greater experience of connectedness, harmony, balance and physical, emotional and psychosocial well-being.

> *HeartMath interventions have facilitated health improvements in patients with:*
>
> - Hypertension
> - Arrhythmias
> - Autoimmune disorders
> - Environmental sensitivity
> - Sleep disorders
> - Drug and alcohol addiction
> - Anger
> - Heart failure
> - Chronic pain
> - Fibromyalgia
> - Chronic fatigue
> - Anxiety disorders
> - Depression
> - PTSD
> - ADD/ADHD
> - Eating disorders

Health-care professionals worldwide, representing both mental health and medical fields, are incorporating HeartMath self-regulation techniques and practices into their treatment strategies with notable success. A growing number of clinical studies and case histories have documented substantial reductions in symptomatology and improvements in clinical status in a wide variety of conditions after a relatively brief time when their patients use these techniques and practices. Collectively, results indicate that such self-regulation techniques are easily learned and employed, produce rapid improvements, have a high rate of compliance, can be sustained over time and are readily adaptable to a wide range of age and demographic groups.

The use of interventions utilizing the HeartMath self-regulation techniques and HRV coherence feedback technology to reduce stress has significantly improved key markers of health and wellness. For example, studies have shown the use of these self-regulation techniques increases parasympathetic activity (HF power) [133] and results in significant reductions in cortisol and increases in DHEA (Figure 8.1) over a 30-day period [116]. In a study for which results are shown in the figure, 30 participants were taught the Cut-Thru and Heart Lock-In self-regulation techniques and practiced using them in daily life for one month. The significant changes in hormonal balance correlated with the significant improvements in emotional health and reductions in stress, anxiety, burnout and guilt along with increases in caring and vigor.

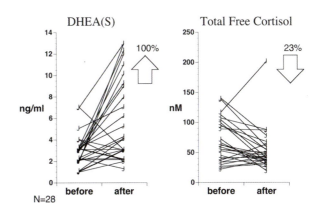

Figure 8.1. DHEA and cortisol values before and after subjects were trained in and practiced HeartMath self-regulation techniques for one month. There was a 100% average increase in DHEA and a 23% decrease in cortisol.

Coherence and Blood Pressure

Several studies have shown significantly lowered blood pressure (BP) and stress measures. Employees with a diagnosis of hypertension who were enrolled in a workplace-based risk-reduction program exhibited significant reductions in blood pressure relative to the control group after using HeartMath tools for three months.[115] Participants also experienced significant reductions in distress and depression, concurrent with improvements in work performance-related parameters following the intervention (Figure 8.2).

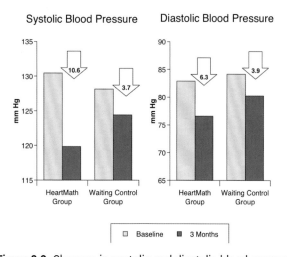

Figure 8.2. Changes in systolic and diastolic blood pressure in the HeartMath group versus the control group. BP was measured before and three months after the completion of the training program. The trained group demonstrated a mean adjusted reduction of 10.6 mm Hg in systolic BP and of 6.3 mm Hg in diastolic BP. (Three-month measurements are adjusted for baseline BP, age, gender, body mass index and medication status.) *p < .05.

The trained group demonstrated a mean adjusted reduction of 10.6 mm Hg in systolic BP and of 6.3 mm Hg in diastolic BP, as compared to reductions of 3.7 mm Hg (systolic) and 3.9 mm Hg (diastolic) in the control group. In addition, three individuals in the trained group were able to reduce their BP medication usage, with their physicians' approval, during the study period. Of these, one participant was permitted to discontinue antihypertensive medication usage entirely following completion of the study. These BP improvements achieved by the treatment group are notable when viewed in comparison to BP reductions typically achieved with other types of interventions. The reduction in BP obtained through the stress-management training is similar in magnitude to the average reduction in BP reported in a meta-analysis of controlled trials of anti-hypertensive drug therapy that lasted for several years. This reduction in BP is the equivalent of a 40-pound weight loss, and is twice the size of the average reduction seen with, for example, a low-salt diet or exercise training.[263-265]

In another study of hypertensive patients, it was found that those who used the techniques to increase HRV coherence had rapid reductions, on average, of 10mm Hg in mean BP. The study was a randomized controlled design with 62 hypertensive participants who were divided into three groups. (Group 1 participants were taking hypertensive medications and taught the Quick Coherence self-regulation technique and used a heart rate variability (HRV) coherence-training device. Group 2 members were not yet taking medications and were trained in the Quick Coherence Technique; those in Group 3 were taking hypertensive medications and did not use the Quick Coherence Technique, but instead were instructed in a relaxation technique that they used between the BP assessments. An ANCOVA (analysis of covariance) was conducted to compare the effectiveness of three different interventions at reducing blood pressure. The two groups that used the Quick Coherence self-regulation technique and HRV coherence-training device were associated with a significantly greater reduction in systolic and mean arterial pressure (average blood pressure) compared with the medication/relaxation technique-only group. The greatest reductions in blood pressure were associated with the combination of medications and the use of the HRV coherence device and the self-regulation technique. Surprisingly, the group that was not taking medications had greater reductions than the medications/relaxation group (Figure 8.3).[113]

Health Risk Reduction in Correctional Officers

A study of 88 California correctional officers with high workplace stress was randomized to experimental and wait-list control groups, stratified on relative health risk, age and gender.[266] The experimental group participated in a stress- and health-risk reduction program, which was delivered over two consecutive days.

Chapter 8: Health Outcome Studies

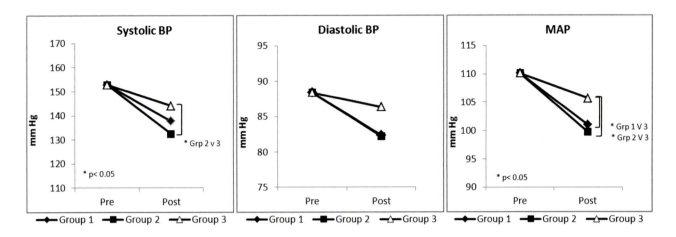

Figure 8.3. Shows the pre-post intervention changes in systolic, diastolic and mean arterial blood pressure for the three groups. The use of the Quick Coherence self-regulation technique and HRV coherence-training device was associated with a significantly greater reduction in systolic and mean arterial pressure (average blood pressure) in the two groups who used the intervention compared with the medication/relaxation technique-only group. The greatest reductions in blood pressure were associated with the combination of medications and the intervention using the HRV device and self-regulation technique.

The program included instruction on health-risk factors as well as training in HeartMath's self-regulation techniques. Learning and practice of the techniques were enhanced by HRV coherence feedback. Physiological changes in the experimental group included significant reductions in total cholesterol, LDL cholesterol levels, the total cholesterol/HDL ratio, fasting glucose levels, mean heart rate, mean arterial pressure, and both systolic and diastolic blood pressure (Figure 8.4).

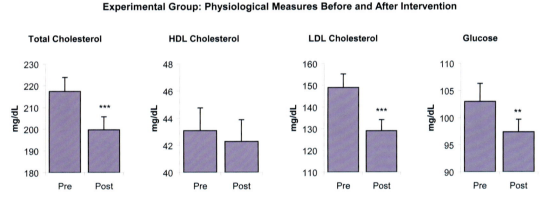

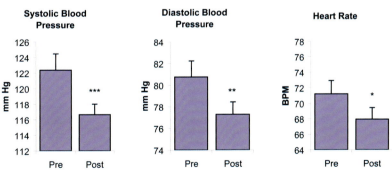

Figure 8.4. Bar graphs illustrate physiological variables in the experimental group, measured before and three months after the intervention program. The group showed significant reductions in total cholesterol, LDL cholesterol, blood glucose levels, systolic and diastolic blood pressure, and heart rate after the intervention. *$p < 0.05$, **$p < 0.01$, ***$p < 0.001$.

© Copyright 2015 HeartMath Institute

Psychological changes included significant reductions in overall psychological distress, anger, fatigue, hostility, interpersonal sensitivity, impatience, and global Type A behavior, and increases in gratitude and positive outlook. There were also improvements in key organizationally relevant measures in the experimental group after the program, including significant increases in productivity, motivation, goal clarity and perceived manager support. Finally, a detailed analysis was performed to calculate the projected health-care cost savings to the organization that would likely result from the reduction in participants' health risk factors. According to this analysis, the reductions in health-risk factors achieved in this study were projected to lead to an average health-care cost savings of $1,179 per employee per year.

Health-Care Cost Reduction

The Reformed Church in America (RCA) identified stress among its clergy as a major cause of higher-than-average health claims. Because the pastors were spread across the U.S., the intervention was provided by a small team of HeartMath certified mentors in six phone sessions to help the participants manage stress and increase physiological resilience. The study divided 313 participants into two groups with 149 participating in the HeartMath program delivered by phone, which included instruction in use of the portable version of the emWave and practice of the self-regulation techniques and 164 in the active control group participated in a phone-based lifestyle-management program. All participants completed a health-risk assessment and a validated HeartMath Stress and Well-being Survey at the beginning of 2007 and again at the beginning of 2008. Well-being, stress management, resilience, and emotional vitality were significantly improved in the HeartMath group compared to the lifestyle-management group. In an analysis of the claims costs data for that year, the pastors who had participated in the HeartMath program were compared to the control group. Adjusted medical costs were reduced by 3.8% for HeartMath participants while there was an increase of 9.0% for the control group. The adjusted pharmacy costs increased 7.9% for the HeartMath group and 13.3% for the control group. The total 2008 savings as a result of the program were $585 per participant, yielding a return on investment of 1.95:1. In the detailed medical cost analysis, one of the higher cost-savings categories was for essential hypertension, which would be expected to be sensitive to reduced stress.[112]

Metabolic Syndrome

A number of significant health outcomes were found in two workplace pilot studies of utility line workers and employees of an online travel company. These studies focused on reducing stress and metabolic syndrome risk factors with the HM self-regulation techniques combined with HRV coherence feedback. In both studies, there were significant reductions in organizational stress (life pressures, relational tensions, work-related stress), emotional stress (anxiety, depression, anger) and stress symptoms (fatigue, sleep, headaches, etc.), and significant increases in emotional vitality. Both studies also showed reductions in the number of participants who were classified as having metabolic syndrome. In the utility-company cohort, total cholesterol and LDL cholesterol were significantly reduced, and the travel-company cohort had significant reductions in both systolic and diastolic BP and triglycerides (unpublished data).

Asthma/Pulmonary Function

One of the reasons coherence training is an effective approach for reducing both short-term and long-term BP, may be a resting of baroreflex gain. Psychophysiologist Paul Lehrer, has shown that using HRV feedback to promote a state of physiological coherence, which he calls "resonance," resulted in lasting increases in baroreflex gain, independent of respiratory and cardiovascular changes.[111] In a large controlled study involving patients with asthma, those using the HRV resonance training had improved lung function, decreased symptoms, exhibited no asthma exacerbations and were able to reduce steroid medications.[267] In other studies, Lehrer demonstrated that pulmonary function improvements occurred in both older and younger patients even though older individuals have lower HRV[183] and that the improvements occur with HRV biofeedback training, but not with relaxed breathing or muscle tension relaxation.[268] He also published a report of 20 case studies which showed uniform improvements in pulmonary function in children with asthma.[269] Additionally, according to Lehrer, in a controlled study, patients with multiple unexplained symptoms and depression[270] showed improvements, as did patients with fibromyalgia[271] and major depression.[272]

Congestive Heart Failure

A study was conducted at Stanford University to evaluate the effect of the HeartMath self-regulation skills training on quality of life and functional capacity in elderly patients with class I–III congestive heart failure (CHF).[192] Thirty-three multiethnic patients (mean age, 66±9 years) were randomly assigned to a treatment group or wait-listed control group. The intervention was provided in eight weekly sessions over a 10-week period. Significant improvements were noted in perceived stress, emotional distress, six-minute walk and depression, and positive trends were noted in each of the other psychosocial measures. The investigators noted that CHF patients were very willing participants and the study suggested that HeartMath techniques were a feasible and effective intervention for CHF patients, demonstrating that stress and depression levels could be reduced and functional capacity increased in this population through training in emotion self-management. This study's promising indications clearly warrant larger-scale controlled trials to confirm the observed psychosocial and functional improvements and further explore the implications of such outcomes for physiological rehabilitation (Figure 8.5-8.7).

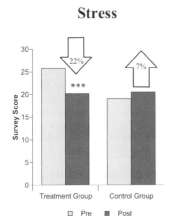

Figure 8.5. Reduction in stress in congestive-heart-failure patients after the HeartMath training program. Stress dropped 22% in the treatment group following the intervention, while it rose 7% in the control group over the three-month study period. (Perceived Stress Scale) ***p < .001.

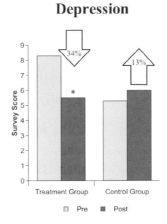

Figure 8.6. Reduction in depression in congestive-heart-failure patients after the HeartMath training program. Depression decreased by 34% in the treatment group whereas it increased by 13% in the control group over the study period. (Geriatric Depression Scale) *p < .05.

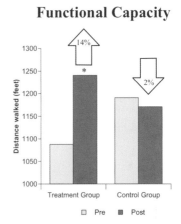

Figure 8.7. Improvements in functional capacity in congestive-heart-failure patients after the HeartMath program. Functional capacity, as measured by performance on the six-minute walk, increased 14% in the treatment group while it declined 2% in the control group. Treatment-group participants were able to walk an average of 153 feet further in six minutes at posttest than at pretest. *p < .05.

Diabetes

In a study of diabetes patients, the introduction of the self-regulation skills led to an improved overall quality of life and glycemic regulation, which correlated with use of the self-regulation techniques.[273] Twenty-two patients with Type 1 or Type 2 diabetes mellitus participated in a two-day training. Hemoglobin A1c, cholesterol and triglycerides, and blood pressure were assessed along with measures of stress, psychological status and quality of life before and six months following the training. There were significant reductions in psychological symptomatology and negative emotions, including anxiety, depression, anger and distress and significant increases in peacefulness, social support and vitality, as well as reductions in somatization, sleeplessness and fatigue.

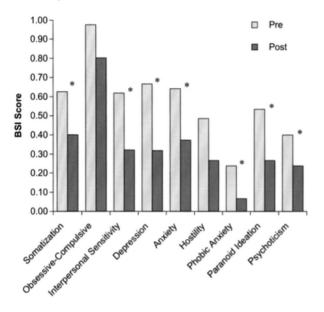

Figure 8.8. Diabetic patients demonstrated significant reductions in a numerous psychological symptoms (Brief Symptom Inventory) after practicing the HeartMath interventions for six months. *p < .05.

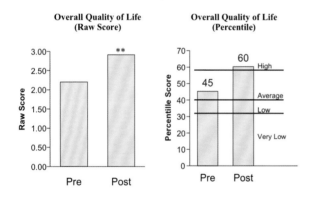

Figure 8.9. The graph on the left illustrates the significant increase in the group's mean overall quality of life raw score, as measured by the Quality of Life Inventory three weeks before versus six months after the HeartMath program. The graph on the right plots the mean overall quality of life percentile score for study participants as compared to normative data. Before the intervention program, the group's mean percentile score plotted very near the bottom of the average range, whereas six months after the program it had moved into the high range. **p < .01.

Participants also showed reduced sensitivity to daily life stressors, and quality of life significantly improved (Figures 8.8 and 9). Regression analysis revealed a significant relationship between self-reported prac- tice of the techniques learned in the program and the change in HbA1c levels in patients with Type 2 diabetes with more practice being associated with reductions in HbA1c.

Coherence and Improved Cognitive Function

Several studies have shown that increased levels of heart-rhythm coherence are associated with significant improvements in cognitive performance.[5, 108, 109] Significant outcomes have been observed in discrimination-and reaction-time experiments and more complex domains of cognitive function, including memory and academic performance.[5, 274] In terms of healthier cognitive and emotional functioning, significant reductions in stress, depression, anxiety, anger, hostility, burnout and fatigue, and increases in caring, contentment, gratitude, peacefulness, resilience and vitality have been measured across diverse populations.[275-280]

Attention Deficit Hyperactivity Disorder (ADHD)

ADHD, the most commonly studied and diagnosed behavioral condition in childhood, is estimated to affect 3% to 5% of children globally. Left untreated, ADHD can lead to academic underachievement, poor interpersonal relationships, anxiety, depression and increased risk of criminal activity. Of particular concern is the increased risk of mental health problems for adolescents with ADHD.

A randomized blind, placebo-controlled study was undertaken with 38 children (aged 9 to 13) with a clinical diagnosis of ADHD in Liverpool, England to assess the potential benefits of HeartMath self-regulation techniques. Learning the skills was supported by HRV coherence feedback training. The placebo control consisted of daily 20-minute, one-on-one sessions with a learning assistant for six weeks. During these sessions, each child was free to build a model of choice from Lego building blocks. Cognitive function was assessed using a comprehensive set of computer-based tests of attention, concentration, vigilance, short-term (working) memory and long-term (episodic) memory before the intervention and again six weeks later. Secondary measures included the Conners' Teachers Rating Scale and the Strengths and Difficulties Questionnaire completed by both children and their teachers. After the post-intervention measures were collected, the control group was provided with the same HeartMath self-regulation skills program. Participants demonstrated significant improvements in various aspects of episodic secondary verbal memory, including delayed word recall and immediate word recall and word recognition (Figure 8.10). Significant improvements in behavior also were found. The results suggest that the intervention offers a physiologically based program to improve cognitive functioning and behaviors in children with ADHD.[108]

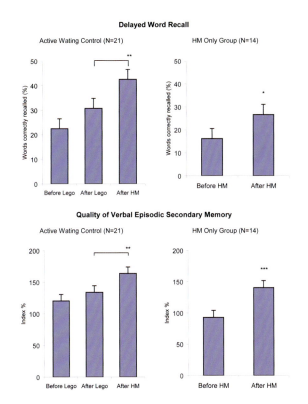

Figure 8.10. Children with ADHD in the HeartMath group exhibited significant increases in overall quality of episodic memory and long-term memory. Children in the placebo-control group (Lego play) showed a small, but non significant improvement over the same time period. There was a significant improvement in the control group as well after they learned and practiced the HeartMath skills. Quality of verbal episodic memory is a composite measure constructed from accuracy measures of four cognitive function tests. *p < 0.05, ** p < 0.01, *** p < 0.01.

Improved Mental Health in Children

A study conducted by the Department of Psychophysiology at an outpatient pediatric clinic in Skopje, Macedonia evaluated the effectiveness of HRV coherence training and the HeartMath self-regulation techniques in the treatment of common mental health disorders in children.[281] Six groups of children were evaluated: a) children with anxious phobic symptoms, N = 15, mean age 12.5 ± 2.25 years; b) children with somatoform problems (somatic symptoms not explained by a general medical condition), N = 15, mean age 10.92 ± 2.06; c) children with obsessive-compulsive manifestations (OCD), N = 7; mean age 14.5 ± 2.20; d) children with ADHD, N = 10, mean age 10.5 ±.1.80; e) children with conduct disorders (CD), N = 12, mean age 11.5 ± 1.52; and f) a control group, N = 15 children, mean age 10.18 ± 1.33. All of the examined children (N = 74) were of similar age. The diagnosis was made according to ICD-10 classification by a team of pediatrician-psychophysiologists, a clinical psychologist and a child neurologist. All children were outpatients at the Skopje pediatric clinic. In the assessment procedure, interviews with parents and children and psychometric evaluations with the Eysenck Personality Questionnaire (EPQ) were used to discriminate four main psychological personality characteristics: extroversion/introversion; neurotic tendencies/stability; psychopathologic traits/normal behavior; and self-control/lie scale.

The biofeedback instrumentation used was HeartMath's HRV coherence-training system (Freeze-Framer, now called emWave Pro). It was used to reinforce the self-regulation techniques. Each patient sat in a comfortable chair in a quiet room with the practitioner. Patients were instructed to practice a self-regulation technique that included rhythmic heart-focused breathing and to activate a positive emotion. After the initial assessment, 15 training sessions were provided. The duration of all sessions was about 16 minutes and included playing two games (Meadow and Balloon), which are controlled by the subject's heart-coherence level.

The results were statistically elaborated with: ANOVA (analysis of variance) for the first and last sessions of all groups and Student t-test for differences between groups. Generally, all children manifesting mental health problems showed lower scores for extroversion and higher scores for neuroticism, compared to the control group at baseline. This finding was considered important in the choice of biofeedback modality. Namely, for introvert personalities, manifesting so-called "inner arousal," the application of peripheral biofeedback modalities was considered to be a better choice. In this context, peripheral biofeedback based on HRV coherence was chosen for the study.

It was found in the results analysis that significantly lower heart rate between the first and last session were obtained for obsessive-compulsive and conduct disorders and anxiety. It means that with training, almost all children, except the ADHD group, learned to lower their heart rate. The Anxiety group showed very good results related to HRV; they were able to increase HRV coherence scores. For children with somatoform problems, there also were significant increases in HRV parameters (VLF, LF and HF power). Changes in HRV in the OCD group also had significant increases in all HRV measures, which were considered to be an important clinical outcome. The children with ADHD did not have increased HRV. The coherence scores increased in all training sessions for all groups, but the highest increases were in the conduct disorder group followed by the general anxiety group (32.5 and 30 respectively). It also improved significantly in the OCD and somatoform disorders groups. HRV training showed very positive results relative to clinical outcomes, especially for children with conduct and anxious-phobic disorders, and for obsessive-compulsive and somatoform disorders.

In general, the authors concluded that HRV coherence training, as a peripheral biofeedback modality, could be a good noninvasive choice, especially for introverted children manifesting common mental health problems, and that the approach has a good cost-benefit ratio. The games included in the training are very engaging for children.

Coherence Training Improves Memory

In a study conducted by Keith Wesnes in London, 18 healthy adult participants (six females, 12 males, ages 20 to 53, mean 32 years) were recruited for a study to assess the potential long-term effects of HeartMath self-regulation skills and HRV coherence training on cognitive performance.[282]

Cognitive function was assessed using a comprehensive set of computer-based tests for attention, concentration, vigilance, short-term (working) memory and long-term (episodic) memory before the intervention and again seven weeks later. Each participant's ECG was recorded for a 10-minute period for HRV and coherence analysis before administration of the cognitive function test battery. In addition, participants completed a short self-administered questionnaire that measured calmness and alertness. After baseline collection, the participants attended a training program in which they learned the Freeze Frame, Heart Lock-In, and Coherent Communication techniques, and instruction in using the HRV coherence feedback system (Freeze-Framer, now called emWave Pro). They were asked to use the Freeze Frame Technique whenever they experienced stress or emotional discord, and the Heart Lock-In Technique three times per week for at least 10 minutes. In addition, they were encouraged to practice the Coherent Communication Technique when engaging in conversation with others. Seven weeks later the participants completed the same measures using exactly the same protocols as were used and followed for baseline data collection.

The results of the pre- and post-analysis of the cognitive performance tests showed significant improvement ($p = 0.0049$) in the quality of episodic (long-term) memory and marginally significant improvement ($p = 0.078$) in the quality of working (short-term) memory (Figure 8.11). There was a positive trend in the composite scores reflecting the ability to pay attention and the speed with which they were able to retrieve information from memory. However, the improvements in these measures did not quite reach statistical significance. Analysis of the questionnaire data showed that the research participants reported feeling significantly calmer at the end of the study than they did at the beginning (t-test 2.44, p < 0.05). This finding is notable, in that Dr. Wesnes reported the magnitude of the improvement was significantly higher than the improvement in quality of memory obtained in a large clinical 14-week trial of the effects of a phytopharmaceutical memory enhancer (a gingko/ginseng combination) on the memory of healthy volunteers.

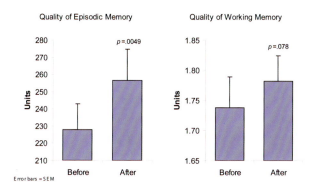

Figure 8.11. Mean improvements in quality of episodic (long-term) memory and quality of working (short-term) memory after participants practiced HeartMath coherence-building tools for seven weeks.

For HRV analysis, standard time and frequency domain HRV measures and coherence levels were computed. In relation to baseline measurement, a significant increase in heart-rhythm coherence ($p < 0.001$) was observed post-intervention before the participants were administered the cognitive function assessments. The group's mean HRV power spectra showing the pre-post differences are shown in Figure 8.12. The increase in power around the 0.1-hertz frequency range indicates a pronounced increase in heart-rhythm coherence, and it occurred even though the participants were *not* specifically instructed to use any of the tools they had learned in the program.

In an effort to explain the observed pre-post changes in the quality of episodic memory and in self-rated calmness, two additional stepwise multiple regressions were run. Of the 10 independent variables included in each analysis, improvement in coherence was the only variable with sufficient statistical power to meet the criterion for entry into the stepwise analysis.

The results show that the change in coherence is quite strongly related to the observed changes in episodic memory and calmness: it accounted for 21% of the variance in the improvement in long-term memory and 42% of the variance in the reported increase in calmness.[5]

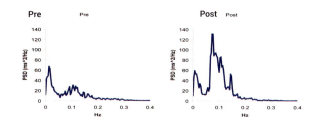

Figure 8.12. Group mean HRV power spectra calculated from 10-minute ECGs recorded before subjects completed the cognitive performance assessments. The left-hand graph shows the mean HRV power spectrum *before* participants were trained in the HeartMath self-regulation techniques, while the right-hand graph shows the mean power spectrum *after* they learned and practiced the techniques for seven weeks. Note the increase in power around the 0.1-hertz frequency range, indicating a pronounced increase in heart-rhythm coherence. This shift is particularly notable, as subjects were not specifically instructed to use the techniques during the post-recording.

Self-Regulation, PTSD Chronic Pain and Brain Injury

While overall health and wellness benefits have been associated with increased coherence, there is also evidence related more specifically to high-stress populations. A study at the William Jennings Bryan Dorn Veterans Affairs Medical Center in Columbia, S.C. of recently returning soldiers from Iraq who were diagnosed with PTSD, found that relatively brief periods of cardiac coherence training combined with practicing the Quick Coherence Technique resulted in significant improvements in the ability to self-regulate and significant improvements in a wide range of cognitive functions, which correlated with increased cardiac coherence. The study also found that resting HRV data in those with a PTSD diagnosis had lower levels of HRV and lower levels of coherence than control subjects without PTSD. Figure 8.13 shows the changes in cognitive function measures, and Figure 8.14 is an example of the typical changes in the HRV waveforms and power spectra from one of the participants.[109]

Science of the Heart

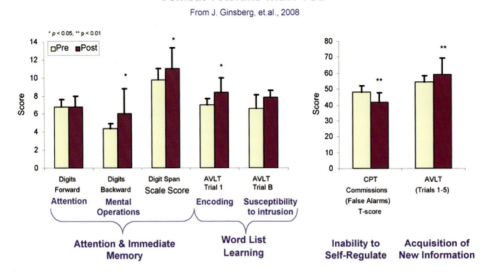

Figure 8.13. Improvements in various measures of cognitive function in recently returned combat veterans with PTSD, after learning the Quick Coherence self-regulation technique and receiving heart-rhythm coherence feedback training using the emWave Pro. Cognitive improvements co-occurred with increases in heart-rhythm coherence and overall HRV. * $p < 0.05$; ** $p < 0.01$.

Figure 8.14 (a and b) depicts the pre-and post-HRVB Training the R-R interval tachogram and power spectra density of one subject with PTSD.

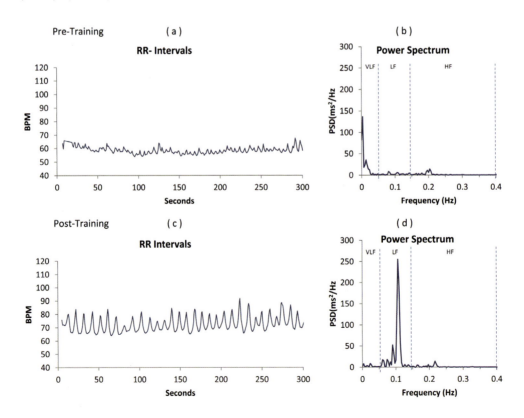

Figure 8.14. Typical example of one of the participant's HRV and power spectra: (a) HRV wave recording pre-training. (b) Power spectrum of HRV. (c) HRV recording post-coherence feedback training. (d) Power spectrum.

In a study of patients with severe brain injury, it was found that the emotion self-regulation training combined with HRV coherence feedback resulted in significantly higher coherence ratios and higher attention scores. Additionally, the families' ratings of participants' emotional control correlated with improved HRV indices.[283]

In a study of returning veterans with chronic pain, pre- and post-measurements of HRV, HRV variables, cardiac coherence, perceived pain, stress, negative emotions and physical activity limitation were made for both treatment and control groups. The treatment group received instruction in the Quick Coherence self-regulation technique, which incorporates controlled breathing and the self-induction of a positive or neutral emotion, along with the HRV coherence feedback device. The technique was practiced during four weekly biofeedback training sessions and was followed by a post-training assessment of pain, stress, and HRV. Control participants simply returned to the lab for a follow-up evaluation four weeks after the initial assessment. The treatment group showed marked and statistically significant increases in coherence (191%) along with significant reductions in pain ratings (36%), stress perception (16%), negative emotions (49%) and physical activity limitations (42%).[284]

Another study conducted in an outpatient pain rehabilitation clinic in a university hospital rehabilitation center in the Netherlands examined the benefits of adding the HeartMath self-regulation and HRV coherence training to a back school (BS) program for patients with chronic, nonspecific low back pain to explore the possible moderators of treatment success.[285] The secondary objective was to test the relationships between HRV coherence at discharge with pain, disability and health perception. It was hypothesized that higher change in coherence scores would be related to changes in pain, disability and health perception.

A total of 170 patients with chronic, nonspecific low back pain were enrolled in the study. Of these, 89 patients were assigned to the standard back school (BS) program and 81 were assigned for BS and heart-coherence training (BS+HCT). Inclusion criteria were: nonspecific lower back pain lasting at least three months and age 18 or older. Exclusion criteria were: mental (e.g. major psychiatric disorders) or physical causes (e.g. cardiac or pulmonary disorders or use of heart medication) or being currently treated for such. A rehabilitation physician approved each participant's inclusion.

At baseline assessment before treatment (T0) and at discharge (T1), the patients filled out a comprehensive set of questionnaires, including demographics, the Pain Disability Index (PDI), the Roland Morris Disability Questionnaire (RMDQ), Numeric Rating Scale (NRS pain) and the RAND 36. HCT was evaluated using a standardized test procedure before HCT and at discharge using a five-minute HRV assessment.

The back school program was provided by experienced physiotherapists on an individual basis. It focused on the physical aspects, such as increase in physical capacity or ergonomics, and behavioral aspects, which could be cognitive behavioral- or acceptance-based approaches. The duration was 12 weeks, two times per week in a cardio fitness setting, for a total of 24 hours in the BS group.

Training in heart coherence and the self-regulation techniques was provided six times, once per week in an individual setting with one hour per meeting, for a total of six hours. Patients were trained in a therapy setting and practiced using the techniques at home. BS-HCT followed a standardized certification program protocol by the HeartMath Institute (HeartMath Interventions Program). The first HCT sessions were focused on learning the basic self-regulation techniques. After the patients learned the basic techniques, they were exposed to stressors by focusing on individualized negative feelings and emotions, including their pain.

Both groups improved significantly on NRS pain, RMDQ, PDI and most of the Rand 36 subscales. On physical functioning, the BS+HCT group improved significantly more than the BS only group (p=0.02). Significant correlations (r=0.39 and r=0.48) were found between increased heart-coherence scores and reduced PDI pain and RMDQ disability scores, but not with other variables. Providing BS-HCT was more effective on physical functioning than the BS-only program was.

Self-Regulation for Caregiver Settings

In recent years, an increasing number of studies on stress have focused on professional caregivers. Affected professionals speak of burnout to define a state of emotional and physical exhaustion caused by the stressful demands of their daily work. The study of these two conditions has been developed in areas of high demand for services such as care for people with dementia. Because dementia is a progressive, disabling and long-term neurodegenerative disease, the risks of being exposed to chronic stress situations for both professional and family-member caregivers is very high.[286] Many studies show the consequences of professional stress, both emotional and physical, that can cause communication problems in the team and families and affect the general welfare of the person.[287] Improving the ability of caregivers to effectively meet the challenges of their daily work is valuable not only for institutions because they will have workers who are less stressed and healthier, but also for people receiving care.[288]

A study conducted in three long-stay nursing homes for elderly people located in different cities in Spain examined the outcomes of providing stress-management intervention based on HeartMath's self-regulation techniques and heart-coherence training with the emWave PSR (now called emWave2) in a group of nursing professionals and family caregivers of elderly patients with dementia.[289] The conceptual nursing model that guided the implementation of this study was Jean Watson's theory of Human Care.[290] Watson's theory emphasizes the importance of taking care of oneself, colleagues or family and others as a means of achieving a more healing environment.

Participants included 42 professionals (67.9% certified nursing assistants) and 32 family-member caregivers. The only exclusion criterion was subjects with sensory or cognitive impairment that prevented them from understanding the training content. A number of socio-demographic variables were collected, including age, sex, occupation, education level, medical history, drug use, years of caregiving and information about people under their care, such as degree of dementia.

The degree of stress and burnout for the professional caregivers was assessed with the Maslach Burnout Inventory (MBI), which has three primary subscales: emotional exhaustion, depersonalization and personal accomplishment at work. For family caregivers, the scale used to assess overload was the validated Spanish version of the Zarit Burden Inventory, which reflects the level of overload a person is experiencing. For measurement of heart coherence, the emWave Pro was used to assess participants' coherence levels at pre- and three-months-post-training during resting state and during a period in which they were asked to relax. In addition, at the end of the workshops, participants' filled out questionnaires on stress and overload, and heart-coherence measures were obtained.

The self-regulation skills training was conducted in workshop style with groups of 10 people, without differentiating between professional and family caregivers. The workshops were provided in one-hour weekly sessions, over a three-month period. Only people who attended more than 80% of the workshops were included in the analysis. The results of an ANOVA analysis of the professional caregivers three months after the training found a significant reduction in the MBI scales for emotional exhaustion and improvement in performance. The depersonalization scale was not significantly changed. In the family caregivers group, the Zarit scale results were not statistically confirmed after a Bonferroni correction (p = .04). A noteworthy finding among family caregivers was a high percentage of hypertension (43.7%), insomnia (28.7%) and anxiety (31.8%), each of which required taking of at least one drug on a daily basis. In professionals, the most prevalent issues were sleep problems (27.8%) and muscular or mechanical (47.6%) problems. Regarding heart-coherence scores, at baseline 58.7% of all participants (n = 71) had low heart-coherence scores. At post assessment, 86.4% of participants had high heart coherence, with significant increases over the baseline values.

The authors concluded that the main objective of this work was to reduce the levels of stress and overload through increased psychological control and increased heart coherence in a group of professionals and caregivers of people with dementia. The results suggest that the intervention achieved the main objective.

Physician Stress Reduction

Given the nature of their occupational duties and environment, physicians often experience work-related stress that may lead to burnout, depression and substance abuse, as well as impaired professional performance. These may be indicated by medical errors, reduced attentiveness or caring behavior toward their patients and other staff members. Physician wellness has been increasingly linked to the quality of patient care, yet the attention that physicians pay to self-wellness often is suboptimal.

A randomized controlled study conducted by Jane Lemarie and colleagues at the University of Calgary with 40 physicians (23 male and 17 female) from various medical practices (1 from primary care, 30 from a medical specialty and 9 from a surgical specialty) was conducted to assess the efficacy of HeartMath self-regulation skills training supported by HRV coherence training (emWave2) for reducing physician stress.[291] Participants in the intervention group were given an HRV coherence training device and were instructed in how to use it. They also participated in an individual training session to learn the Quick Coherence self-regulation technique and were instructed to use the HRV device during the study for five minutes at least three times daily. A research assistant contacted each participant in the intervention group twice weekly to measure stress, heart rate, blood pressure and overall well-being to document adherence to using the stress-management techniques and to record a three-minute HRV session using the emWave Pro system. Participants in the control group received a brochure describing the provincial physician wellness support program and were contacted twice weekly by a research assistant to measure stress, heart rate, blood pressure and overall well-being.

The primary outcome, stress, was assessed with a multiple-item scale developed by the research team, which measured global perceptions of stress and captured occupation-specific stress that is particularly relevant to physicians. The survey included 15 items from the Perceived Stress Scale and 25 items from the Personal and Organizational Quality Assessment–Revised (POQA-R) questionnaire for anxiety, anger, physical symptoms of stress and work-related time pressures. The final 40-item instrument was validated through confirmatory common factor analysis. Pre-intervention data was compared to data collected 28 and 56 days later.

The analysis of the data at day 28 showed that the mean stress score declined significantly for the intervention group (change −14.7, standard deviation [SD] 23.8; p = 0.013) but not for the control group (change −2.2, SD 8.4; p = 0.30). The difference in mean score change between the groups was 12.5 (p = 0.048). The lower mean stress scores in the intervention group were maintained during the trial extension to day 56. After the assessment at day 28 the control group received the same intervention, after which mean stress scores were significantly lower at day 56 (change −8.5, SD 7.6; p < 0.001).

The authors concluded that HRV coherence training along with practice of a self-regulation technique may be a simple and effective stress-reduction strategy for physicians.

Medical Error Reductions

Another study conducted with a large chain of retail stores that had in-store pharmacies employing 220 pharmacists across multiple locations found a reduction in medical errors ranging from 40% to 71%, depending on the store location.[292]

CHAPTER 9

Outcome Studies in Education

Growing evidence of the tremendous benefits to be gained from learning to self-regulate emotions and stress at an early age is becoming increasingly apparent. In today's fast-paced society there is mounting pressure on children to achieve and excel in school at younger and younger ages. Today's children, however, experience considerably greater stress in their lives, shouldering far greater responsibilities and emotional burdens than youngsters their age did even as few as 10 years ago. Many are part of deteriorating families or households in which parents are rarely home and the responsibility for their and their younger siblings' care has fallen largely to them. The majority of these children find little more comfort or security at school, where they often fear becoming victims of bullying and violence and feel pressured to engage in sex or consume drugs and alcohol. Increasing media reports of extreme episodes of violence in schools recently have raised public awareness of children's deteriorating emotional health and underscored the need for more effective solutions to resolve these issues.

"We are educated in school that practice precedes effectiveness, whether in reading, writing, computers, or whatever. We are rarely taught how to practice care, compassion, appreciation or love —essential for family balance."
– Doc Childre

Our educational systems continue to focus on honing children's cognitive skills from the moment they enter the kindergarten classroom. Virtually no emphasis is placed on educating children on managing the inner conflicts and unbalanced emotions they bring with them to school each day. As new concepts such as "social and emotional intelligence" become more widely applied and understood, more educators are realizing that cognitive ability is not the sole or necessarily the most critical determinant of young people's aptitude for flourishing in today's society. Proficiency in emotional management, conflict resolution, communication and interpersonal skills is essential for children to develop inner self-security and the ability to effectively deal with the pressures and obstacles that will inevitably arise in their lives. Moreover, increasing evidence is illuminating the link between emotional balance and cognitive performance. Growing numbers of teachers agree that children come to school with so many problems that it is difficult for them to focus on complex mental tasks and the intake of new information, skills that are essential for effective learning. A substantial body of evidence exists that clearly shows when children also learn social and emotional skills, lifelong benefits that cross domains and expand the mind's capacities can be obtained.

"Some students came to me having memorized the definition of peace, for instance, and they had no idea what it really meant – especially for them personally."
–Edie Fritz, Ed.D., educational psychologist

- In 1940, the top problems in American public schools, according to teachers, were: talking out of turn, chewing gum, making noise, running in the halls and littering. In 1990, teachers identified the top problems as drug abuse, alcohol abuse, pregnancy, suicide and robbery and assault.[293]

- Since 1978, assaults on teachers have risen 700%.[294]

- One youth in six, between the ages of 10 and 17, has seen or knows someone who has been shot.[295]

- A study found that in a group of neglected children, the cortex, or thinking part of the brain, was 20% smaller on average than in a control group.[296]

- Positive emotions have been found to produce faster learning and improved intellectual performance. *B. Fredrickson. Rev Gen Psychol. 1998; 2(3)*

- In a sample of youth ages 7 to 11 years old in the Pittsburgh, Pa. area, over 20% were determined to have a psychiatric disorder.[297]

- Only 37% of youth report feeling a sense of personal power, and half feel that their life has a purpose.[298]

- Since 1960 the rate at which teenagers commit suicide has more than tripled. Suicide is now the second leading cause of death among adolescents.[299]

- The more teenagers feel loved by their parents and comfortable in their schools, the less likely they are to have early sex, smoke, abuse alcohol or drugs or commit violence or suicide.[300]

Self-Regulation and Reduced Test Anxiety and Increased Test Scores

A grant from the U.S. Department of Education provided funding for a randomized controlled study of 980 10th-grade students in two large high schools. The TestEdge National Demonstration Study (TENDS) was conducted by researchers at HeartMath Institute in collaboration with faculty and graduate students in Claremont Graduate University's School of Educational Studies. The study's primary purpose was to investigate the efficacy of the TestEdge program in reducing stress and test anxiety and improving emotional well-being, psychosocial functioning and academic performance in public school students. This involved determining the magnitude, correlates and consequences of stress and test anxiety in a large sample of students and investigating the degree to which the TestEdge program could benefit students in an experimental group, compared to those in a control group. A second programmatic purpose was to characterize the implementation of the program in relation to its receptivity, coordination and administration in a wide variety of school systems with diverse cultural, administrative and situational characteristics.[110, 301]

The study tested two major hypotheses. The first is that competence in the emotion self-regulation skills taught in the TestEdge program would result in significant reductions in test anxiety, which, in turn, would generate a corresponding improvement in academic and test performance. Secondly, as a result of the improvement in student self-regulation skills, it was hypothesized there would be improvements in stress management, emotional stability, relationships, overall student well-being, and in classroom climate, organization and function. To investigate these hypotheses, two studies were conducted, each with different research objectives and designs.

Primary Study

The primary study focused on an in-depth investigation of the entire 10th-grade populations at two large California high schools. One high school was randomly selected as the intervention school, while the other served as the control school. This was designed as a quasi-experimental, longitudinal field study involving pre- and post-intervention measures within a multimethods framework. Extensive quantitative and qualitative data were gathered using a questionnaire developed and validated for the study, interviews and structured observation, and student test scores from California standardized tests – the California High School Exit Examination (CAHSEE) and the California Standards Test (CST). Altogether, 980 students participated in the primary study, of which 636 (53% male, 47% female) were in the experimental group and 344 (40% male, 60% female) were in the control group.

The TestEdge program was taught by English teachers over one semester for approximately four months. In the program, students learned and practiced specific emotion-management techniques to aid them in more effectively handling stress and challenges, both at school and in their personal lives. They also were taught how to apply these techniques to enhance various aspects of the learning process, including test preparation and test-taking. Both the student and teacher programs included use of the Freeze-Framer (now emWave Pro) technology, a heart-rhythm coherence feedback system designed to facilitate acquisition and internalization of the self-regulation skills that were taught.

Across the whole sample at baseline, 61% of all students reported (Spielberger test-anxiety inventory) being affected by test anxiety, with 26% experiencing high levels of test anxiety often or most of the time. Twice as many females as males experienced high levels of test anxiety. There was a strong negative relationship between test anxiety and test performance; students with high levels of test anxiety scored, on average, 15 points lower on standardized tests in both mathematics and English-language arts (ELA) than students with low test anxiety (Figure 9.1). A multiple regression analysis found that *affective mood* measures accounted for the wide variance between student test performance on both the CST and CAHSEE English-language arts exams and Test Anxiety Inventory – global scale (23% versus ~13%, respectively). Positive feelings and prosocial behaviors were found to have positive effects on test performance, while strongly negative feelings and antisocial behaviors had negative effects.[302] Taken as a whole, these findings are sobering and justify the concern that student test anxiety and emotional stress may significantly jeopardize assessment validity and therefore may constitute a major source of test bias.

There was a significant reduction in the mean level of test anxiety. Of those students at the intervention school who had reported being affected by test anxiety at the beginning of the study, 75% had reduced levels of test anxiety by the end of the study. This reduction in test anxiety also was evident in more than three-quarters of all classrooms, and it was observed throughout the academic ability spectrum, from high test-performing classes to low test-performing classes.

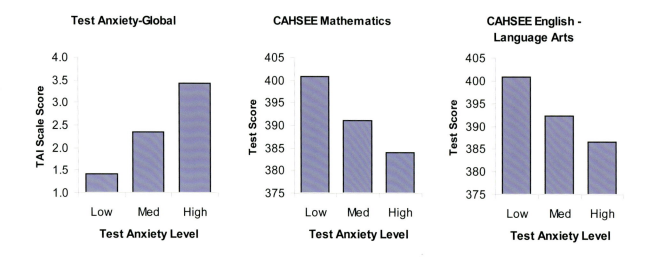

Figure 9.1. Baseline test anxiety, measured by the Test Anxiety Inventory (TAI) – global scale score and California High School Exit Examination (CAHSEE) scores in English-language arts and mathematics have been classified into three approximately equal-sized groupings of students with low, medium, and high test anxiety scores. A strong, statistically significant ($p < 0.001$) negative relationship is clearly apparent between mean level of test-anxiety and mean performance on the standardized tests: As test anxiety increases, test performance decreases.

After the TestEdge program was provided to the students in the intervention school, there was strong, consistent evidence of a positive effect from the intervention on these students, compared to those in the control school. The reduction in test anxiety was associated with significant improvements in social and emotional measures (Figure 9.2), including reductions in *negative affect, emotional discord* and *interactional difficulty*, and an increase in *positive class experience*. In four matched-group comparisons (involving subsamples of 50 to 129 students per grouping), there was a significant increase in test performance in the experimental group over the control group, ranging on average from 10 to 25 points. In two of these matched-group comparisons, this significant increase in test performance was associated with a significant decrease in test anxiety in the experimental group (Figure 9.3).

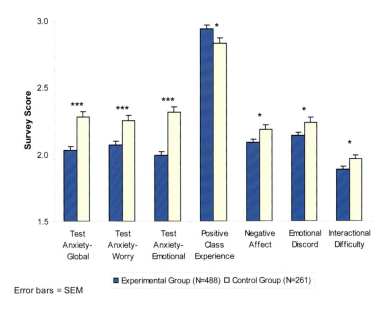

Figure 9.2. Results of an ANCOVA of pre– and post-intervention changes in measures of test anxiety (*global scale, worry component,* and *emotionality component*) and social and emotional scales (*positive class experience, negative affect, emotional discord,* and *interactional difficulty*) showing significant differences between the intervention and control schools. *$p < 0.05$, ***$p < 0.001$.

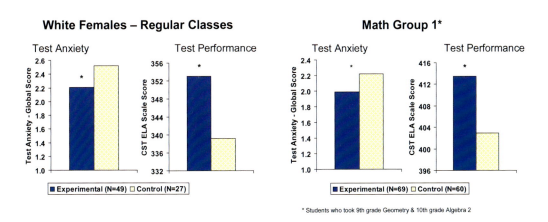

Figure 9.3. ANCOVA results for two subsamples from the intervention and control schools matched on sociodemographic factors (White females in average academic-level classes) and ninth-grade math-test performance (Math Group 1), respectively. For these matched-group comparisons, significant reductions in test anxiety in conjunction with significant improvements in test performance (California Standards Test – English-language arts) were observed in the experimental group, compared to the control group. *$p < 0.05$.

© Copyright 2015 HeartMath Institute

> *"As annual testing becomes a regular part of our educational system through the No Child Left Behind Act and other state requirements, it is important that when we test students on their mastery of a subject, we are truly getting accurate results. The program developed by HeartMath Institute to reduce test-related anxiety shows great potential for helping students achieve success."*
> — Rep. Ralph Regula (R-OH)

Student participants describe how they apply HeartMath skills in various areas of their lives:

"I use HeartMath every day before math class. This helps me understand the concepts and grasp them more easily."

"HeartMath has obviously helped in my algebra II class. On the last test, using HeartMath, I scored a 97%. It really calmed me down and opened my mind to clearer thinking."

"Working at Coney Island, I'm around people all the time; being an assistant manager, the questions, concerns and complaints all come to me. In order to answer them or handle the situation without taking it personally, I use HeartMath. It's helped calm my nerves a lot."

"Sometimes during sports activities I get mad at myself and do very badly. One time I used HeartMath and I let all of my stress go and just had fun. In the end, I performed better than I expected."

"I have used HeartMath during band. I hate the feeling of having to play a hard part in front of the whole band by myself. I get so nervous sometimes that my hands shake and my mind goes blank. ... Using HeartMath, I have seen a significant improvement in my playing alone in front of people."

"I use HeartMath during any stressful encounter I have, whether at home, school, with my parents or with friends. Anytime I get frustrated or upset, I try to practice (HeartMath techniques) and focus from head to heart."

"(One) time, I was taking a test in Chemistry, and I could not remember the equation for the test. I had three minutes left and I used one of those minutes to do HeartMath. ... I ended up getting a 90% on the test and brought my chem grade up."

"(One) instance was a presentation in English class. I am a shy, solitary person and I sometimes get very stressed with presentations. However, I used what I learned from HeartMath and was able to present my topic efficiently."

"Whenever I email or write my boyfriend, who recently left to work in Utah over the summer, I use HeartMath to calm down and say what I want to say without sounding angry, upset – and (not) panic."

Physiological Study Findings

In addition, a physiological substudy was conducted on a randomly stratified sample of students from both schools. Utilizing measures of heart rate variability (HRV), this controlled laboratory experiment investigated the degree to which students had learned the skills taught in the TestEdge program using an objective measurement of their ability to shift into the physiological coherence state before taking a stressful test.[110]

In a controlled experiment simulating a stressful testing situation, students from the intervention and control schools (N = 136) completed a computerized version of the Stroop color-word conflict test (a standard protocol used to induce psychological stress), while continuous heart rate variability recordings were gathered. For the pre-intervention administration of the experiment, after a resting HRV baseline had been collected, students were instructed to mentally and emotionally prepare themselves to perform an upcoming challenging test and activity, after which they participated in the Stroop test. They also were told if they performed well on the test, they would be given a free movie pass. The same procedures were used in the post-intervention assessment.

Results from the post-intervention physiological experiment:

> The HRV data indicated students who had received the TestEdge program had learned how to better manage their emotions and to self-activate the physiological coherence state under stressful conditions during the stress-preparation segment of the protocol (Figure 9.4).

> The ability to self-activate coherence noted above was associated with significant reductions in test anxiety and corresponding improvements in measures of emotional disposition.

> For students matched on baseline test scores, the capacity to self-activate coherence was associated with a reduction in test anxiety as well as an improvement in test scores in the experimental group (Figure 5.1). This finding is consistent with the results for students in the larger study.

> Students in the experimental group also exhibited increased heart rate variability and heart-rhythm coherence *during the resting baseline period* in the post-intervention experiment – even *without* conscious use of the self-regulation techniques. This suggests that through consistent use of the coherence-building tools over the study period, these students had internalized their benefits, thus instantiating a healthier, more harmonious and more adaptive pattern of psychophysiological functioning as a new set-point or norm.

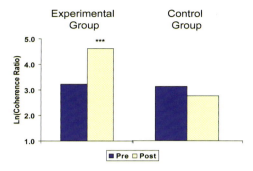

Figure 9.4. These data are from the physiological study, a controlled experiment involving a random stratified sample of students from the intervention and control schools (N = 50 and 48, respectively). These graphs quantify heart-rhythm coherence, the key marker of the psychophysiological coherence state, during the stress-preparation phase of the protocol. Data are shown from recordings collected before and after the TestEdge program. The experimental group demonstrated a significant increase in heart-rhythm coherence in the post-intervention recording when they used one of the self-regulation techniques to prepare for the upcoming stressful test – compared to the control group. ***$p < 0.001$.

Qualitative Findings:

To supplement the quantitative data, the study gathered observations of student classroom interactions in the two schools and conducted structured interviews with teachers. The pre- and post-observational findings were broadly consistent with the findings from the quantitative analysis:

> More positive changes were observed in the social and emotional environment and interaction patterns in the classrooms of the experimental school, while more negative changes were observed in the control school over the course of the semester.

> Students at the experimental school exhibited reduced levels of fear, frustration and impulsivity. They also exhibited increased engagement in class activities, emotional bonding, humor, persistence and empathetic listening and understanding.

> Most teachers acknowledged in interviews that their students came to school emotionally unprepared to learn, but they felt their own educational

training did not equip them with the requisite skills to effectively manage their personal stress or to help their students manage their stress. Teachers were supportive of integrating emotion-management instruction into educational curricula. Most reported experiencing personal benefits as well as observing positive changes in their students' behavior as a result of the intervention program.

Secondary Study

The secondary study consisted of a series of qualitative investigations to evaluate the accessibility, receptivity, coordination and administration of the program across elementary, middle and high schools and in school systems with diverse ethnic, cultural, socioeconomic, administrative and situational characteristics. We employed a case-study approach to evaluate the implementation of the TestEdge program in nine schools in eight states (California, Delaware, Florida, Ohio, Maryland, Texas, Wisconsin, and Pennsylvania). Age-appropriate versions of the TestEdge program were delivered to selected classrooms, covering the third through eighth and 10th grades.

Observational and interview data were gathered to provide information on best practices and potential difficulties when implementing interventions such as TestEdge in widely diverse school settings.

Major Findings from the Secondary Study

Evaluation of the implementation case studies of the TestEdge program, conducted in selected classrooms at various grade levels across different states, produced a number of notable results and largely corroborated the findings from the primary study:

> In teacher interviews, the lack of emotion- and self-management education for students was seen as a significant obstacle to learning and academic performance. Most teachers described positive changes in students' attitudes, behaviors, test anxiety and academic performance, and attributed these to the TestEdge program. They also felt that the tools and skills that were learned would have a positive impact on their students' future social, emotional and academic development.

> Most teachers reported that the program provided substantial benefits in their personal and professional lives.

> In general, the program's implementation was more successful when there were several teachers at the same grade level teaching it and when teachers were able to internalize the use of the tools in their own lives.

> Major challenges to successful program implementation included inadequate class time; logistical problems encountered with school administration; and securing the support of the principal and other school administrators to foster teacher commitment.

Elementary School Case Study:

A case example of highly successful implementation of the TestEdge program was provided by an in-depth study conducted at the third-grade level in a Southern California elementary school. Several notable findings emerged from the study:

> Large increases in state-mandated test scores were observed, which far exceeded academic targets for the year. As a result, student proficiency grew from 26% to 47% in English-language arts and from 60% to 71% in mathematics.

> Corresponding emotional and behavioral improvements also were observed among students in the classrooms.

> The success of implementation was largely the result of the enthusiastic support provided by the school's principal and key teachers and administrators.

Conclusion

Overall, the evidence from this rich combination of physiological, quantitative, and qualitative data indicates that the self-regulation skills and practices taught in the TestEdge program led to a number of important successes. It is our hope that the results of this research will have an impact on policies regarding the importance of integrating stress and emotion self-management education into school curricula for students of all ages. By introducing and sustaining appropriate programs and strategies, it should be possible to significantly reduce the stress and anxiety that impede student performance, undermine teacher-student relationships and cause physiological and emotional harm. Such programs have the promise of increasing the effectiveness of our educational system and, in the long-term, boosting the academic standing of the United States in the international community.

Evaluation of HeartMath Program with Schoolchildren in West Belfast

A study was conducted in Belfast, Ireland to investigate the efficacy of a HeartMath program as a means of improving student emotion self-regulation and associated improvements in emotional stability, relationships and overall student well-being.[303] Seven schools were chosen for the study: three were primary schools (n = 122) and four were post-primary schools (n = 121). Two or three classes from each school participated in the study. Each class participated in a one-day interactive program in which they learned how to use the HRV coherence system (emWave Pro) and was taught an emotion self-regulation technique. The students and teachers also were provided with an audio CD called "Journey to my Safe Place" to lead them through a self-regulation exercise at later times. Teachers were encouraged to allow the students to use the games on the emWave once or twice weekly, as time permitted. The outcome measure was the Strengths and Difficulties Questionnaires (SDQ) for ages 4 to 16; these were completed pre- and post-intervention by the teachers.

Based on the SDQ scales, the primary students showed a statistically significant reduction in emotional problems (51%); conduct problems (43%); hyperactivity (40%); and a significant improvement in relating to peers (50%). According to the scales, post-primary students had significant reductions in hyperactivity (12%), emotional problems (9%) and conduct problems (9%), and they showed (27%) improvement in relating to peers. As expected, the primary children benefitted more than post-primary students, and the authors suggest this may be the result of better adherence to utilizing the skills, which are easier to reinforce in primary-school students because they are in the same classroom each day. In contrast, the students in the post-primary schools move from class to class throughout the day, so it was more difficult to set up a routine to support the intervention. In addition, most of the primary-school classes began the study several weeks before the post-primary classes, which gave them more time to plan how the program would be integrated into their classes.

Self-Regulation for Promoting Development in Preschool Children

The importance of children learning effective social and emotional skills at an early age cannot be overestimated. A large body of research clearly shows that learning how to process and self-regulate emotional experience in infancy and early childhood not only facilitates neurological growth, but also determines the potential for subsequent psychosocial and cognitive development. Conversely, the inability to appropriately self-regulate feelings and emotions can have devastating long-term consequences in a child's life, impeding development and often resulting in impulsive and aggressive behavior, attentional and learning difficulties and an inability to engage in prosocial relationships. Furthermore, this early deficiency is associated with later-forming psychosocial dysfunction and pathology that not only robs individuals of a fulfilling life, but also results in an enormous cost to our society.[80]

The HeartMath Institute developed a program called Early HeartSmarts® (EHS) specifically intended to help equip children aged 3 to 6 with the foundational emotion self-regulation and social competencies for school. The program trains teachers to guide children in learning emotion self-regulation and key, age-appropriate social and emotional skills, toward a goal of promoting children's emotional, social, and cognitive development. The Early HeartSmarts program included interactive activities, games and self-management tools designed to promote children's learning of several key emotional and social competencies, which in turn are known to promote psychosocial development. These included:

- How to recognize and better understand basic emotional states.
- How to self-regulate emotions.
- Ways to strengthen the expression of positive feelings.
- Ways to improve peer relations.
- Skills for developing problem-solving.

A study was conducted in the Salt Lake City School District to evaluate the efficacy of the Early HeartSmarts program.[304] The study was conducted over the school year using a quasi-experimental longitudinal field research design with three measurement moments (baseline and pre- and post-intervention panels). Children in 19 preschool classrooms at 19 schools were divided into intervention and control group samples (N = 66 and 309, respectively; mean age = 3.6 years). Classes in the intervention group were selected by the district to target children of lower socioeconomic and ethnic minority backgrounds. The Creative Curriculum Assessment (TCCA), a teacher-scored, 50-item instrument, was used to measure student growth in four areas of development: social/emotional, physical, cognitive and language development.

Teachers delivered the program to their students throughout the second half of the school year. A key element of the program was learning and practicing two simple HeartMath emotion-shifting tools: Shift and Shine and Heart Warmer.

Overall, results provided compelling evidence of the efficacy of the EHS program increasing total psychosocial development and each of the four development areas measured by the Creative Curriculum Assessment: The results of a series of ANCOVAs found a strong, consistent pattern of large, significant differences on the development measures favoring preschool children who received the EHS program over those in the control group (see Figure 9.5). From the adjusted means on the five development scales, a significant difference with a large effect size was observed favoring the intervention group on the total development scale (ES 0.81, $p < 0.001$), and on each of the social/emotional development (ES 0.97, $p < 0.001$), physical development (ES 0.79, $p < 0.001$), cognitive development (ES 0.55, $p < 0.01$), and language development (ES 0.73 $p < 0.001$) scales. And at the subcomponent level (right-hand graph in Figure 9.5), the intervention group demonstrated a statistically significant improvement in all 10 constructs, eight of which showed differences that were large in terms of effect size. The magnitude of development observed for the intervention-group children is particularly striking, considering that initial measurements indicated that at the beginning of the study, they started with a significant development handicap relative to their peers in the control group. After participating in the EHS program, they had surpassed the control group's development growth by the end of the study. A further important finding, from an analysis of demographic factors, was that the program was effective in promoting development in children across all the sociodemographic categories investigated: males, females, Hispanic, Caucasian, free lunch (an indicator of low socioeconomic status) and no free lunch.

An important point to emphasize is that these results are for preschool children, who are *very* young, and 96% of them were 3.0 to 4.0 years old. It is both striking and remarkable that children as young as 3 can begin to learn and practice the emotion self-regulation skills, which appears to promote their development across a wide range of categories. Given that the age range from 3 to 6 years is a period of accelerated neurological growth and development, it is likely that the learning and sustained use of these skills and practices during this important developmental period will readily instantiate a new set point in a young child's nervous system for an optimal pattern of emotion self-regulation and healthy social function, thereby significantly boosting the development trajectory of future prosocial behaviors and academic achievement.

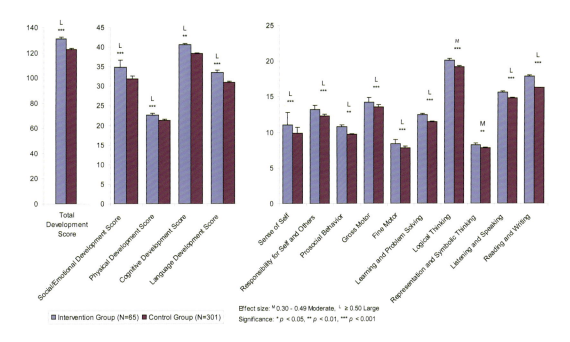

Figure 9.5. Adjusted means showing results of ANCOVA of intervention effects on development measures comparing intervention and control groups.

Reduced Burnout in College Students

Dr. Ross May and colleagues at Florida State University have found that the psychophysiological functioning underlying school burnout is of particular importance. Their research has shown student burnout is associated with increased markers of cardiovascular risk and poor academic performance (GPA).[305, 306] They suggest that because school burnout is associated with increased cardiac risk, it should be recognized as a potential public health issue and a cause for concern for university officials. This is especially true because cardiovascular disease (CVD), including hypertension, coronary heart disease, heart failure, peripheral artery disease and stroke, is the most salient cause of death in the United States and the rest of the world. The cardiovascular responses seen from individuals suffering from higher levels of burnout have been identified as risk factors for the future development of cardiovascular disease.[307-309] They also have demonstrated that school burnout is a stronger predictor of GPA than anxiety and depression.

May and his colleagues conducted a study comparing the effects of training in HeartMath self-regulation techniques, supported by HRV coherence training, with high-intensity aerobic training (HIIT); both types of training were intended to ameliorate school burnout in undergraduate students. A total of 90 participants (freshman year, mean age = 18.55, SD = 0.99, 82% female) were randomly assigned to one of the following three groups: one that received the HeartMath Building Personal Resilience program, which included learning self-regulation techniques and HeartMath's computer-based HRV coherence training devices (emWave), high-intensity aerobic training and a no-intervention control. All of the groups were evaluated for cognitive, psychological and cardiovascular functioning before and after a four-week intervention period. The ethnic composition of the sample was 70% Caucasian, 7% African American, 13% Hispanic, 7% Asian and 3% undisclosed ethnicity.

All participants completed a physical health history questionnaire, the School Burnout Inventory (SBI), Center for Epidemiologic Studies Depression Scale (CES-D), and the State-Trait Anxiety Inventory (STAI), and a self-reported sleep quality questionnaire. In addition, academic absenteeism (how many classes were missed over the semester) and GPA were assessed. Cognitive functioning was assessed with computerized working-memory span measure versions of the common reading and operation working memory span tasks.[310, 311] The span tasks require participants to remember target letters while performing a concurrent reading comprehension (reading span) or arithmetic task (operation span). The number of targets in a trial set varied between two and five, with three trials of each size for each of the tests. Physical fitness with a stationary bike test and VO_2 max and cardiovascular functioning were assessed by measures of aortic hemodynamics, beat-to-beat blood pressure, and heart rate variability.

The HeartMath resilience training and HRV coherence sessions were conducted at the university wellness center by trained student instructors three times per week over a four-week period. Each student practiced shifting and sustaining heart-rhythm coherence while using an emWave device and was encouraged to use the techniques and device on a regular basis to help them improve their self-regulation skills and physiological and psychological balance. The high-intensity interval training (HIIT) comprises brief bursts of intense exercise separated by short periods of recovery. This method of exercise is a time-efficient stimulus to induce physiological adaptations normally associated with continuous moderate-intensity training.[312] As few as six sessions of HIIT over two weeks has been shown to increase muscle oxidative capacity to the same extent as a continuous moderate-intensity training protocol that requires an approximately threefold greater time commitment and approximately ninefold higher training volume.[313] The HIIT training sessions were conducted at the university wellness center by trained instructors three times per week over four weeks. The control group visited the wellness center to report their normal daily activities for the duration of the study. Participants were encouraged to not change their normal daily routines.

In comparison to HIIT and control, the HeartMath group participants had significant improvements in academic success and concentration and significant decreases in test anxiety and absenteeism (Figure 9.6).

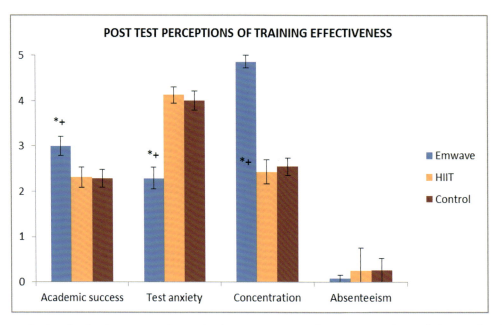

Figure 9.6 shows the data for the three groups for academic success, test anxiety, concentration and absenteeism from classes over the semester. Data are mean and 95% CI. * = p <.05 HeartMath vs. HIIT posttest, + = p <.05 HeartMath vs. Control, a = p <.05 HeartMath pretest vs. HeartMath posttest, b = p <.05 HIIT pretest vs. HIIT posttest, c = p <.05 Control pretest vs. Control posttest.

The HeartMath group also was the only group to show a significant reduction in school burnout from the pretests to posttests (Figure 9.7) and significant improvements in the cognitive functioning assessments of reading and operational span working memory capacity from the pretests to posttests (Figure 9.8 and 9).

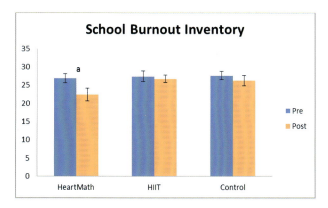

Figure 9.7 shows the pre- and post-data for the three groups for school burnout. Data are mean and 95% CI a = p <.05 HeartMath pretest vs. HeartMath posttest.

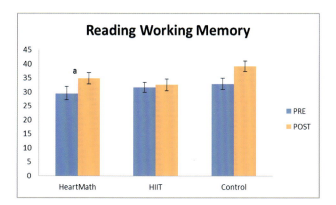

Figure 9.8 shows the pretest and posttest data for the three groups for reading working memory. Data are mean and 95% CI a = p <.05 HeartMath pretest vs. HeartMath posttest.

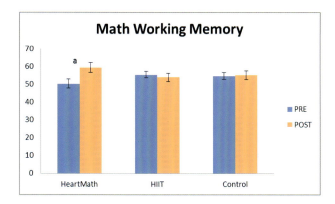

Figure 9.9 shows the pretest and posttest data for the three groups for math working memory. Data are mean and 95% CI a = p <.05 HeartMath pretest vs. HeartMath posttest.

In terms of cardiovascular functioning, the HeartMath group showed significantly decreased brachial and aortic blood pressure, and both the HeartMath and HIIT groups had significantly decreased heart rate from pretest to posttest (Figure 9.10, 11 and 12). The HRV metrics showed that the normalized LF power was significantly lower in the post-assessment measures, compared to the pre-intervention measures, and HF power was significantly increased in all three groups (Figures 9.13 and 14). In addition, the HIIT group was the only group to show significantly improved VO2 max.

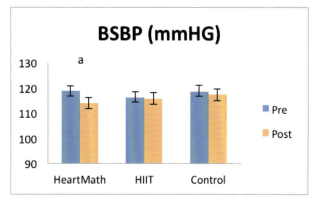

Figure 9.10 Brachial systolic blood pressure (BSBP) data for the three groups. Data are mean and 95% CI. a = p <.05 HeartMath pretest vs. HeartMath posttest.

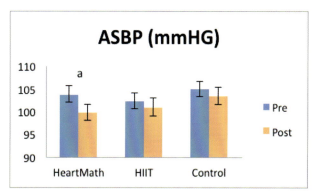

Figure 9.11 Aortic systolic blood pressure (ASBP) data for the three groups. Data are mean and 95% CI a = p <.05 HeartMath pretest vs. HeartMath posttest.

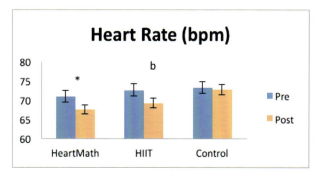

Figure 9.12 Heart-rate data for the three groups. Data are mean and 95% CI. * = p <.05 HeartMath vs. HIIT, b = p <.05 HIIT pretest vs. HIIT posttest.

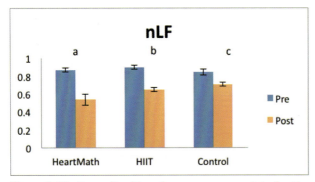

Figure 9.13 Normalized low frequency power data for the three groups. Data are mean and 95% CI. [a] = p <.05 HeartMath pretest vs. HeartMath posttest, [b] = p <.05 HIIT pretest vs. HIIT posttest, [c] = p <.05 Control pretest vs. Control posttest.

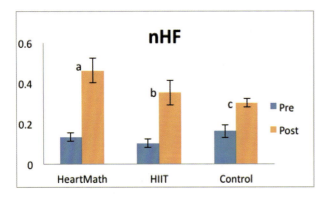

Figure 9.14 Normalized high frequency power data for the three groups. Data are mean and 95% CI. [a] = p <.05 HeartMath pretest vs. HeartMath posttest, [b] = p <.05 HIIT pretest vs. HIIT posttest, [c] = p <.05 Control pretest vs. Control posttest.

Coherent Learning: Creating High-level Performance and Cultural Empathy From Student to Expert

Nursing schools are charged with graduating nursing students who reflect the race and ethnicity of the communities those schools serve. In 1998, 19 Native American students were admitted to the University of Oklahoma College of Nursing. Only 12 had graduated two years later. The rate of attrition for Native American nursing students averaged 57% between 1997 and 2001, compared to the overall attrition rate of approximately 9%.

The OU College of Nursing identified staff members to become certified in a HeartMath program in 2002. The program was implemented starting in 2003. Participation in the program was voluntary for the first year, but it became part of the new-student ori-

entation the next year. Training was offered monthly for students and faculty and was available to every student. Laboratory computers were equipped with the HRV coherence training technology to support the self-regulation skills so students could practice during school hours. Several faculty members also provided student mentoring and practiced with students in their offices at students' request. Many faculty members gave short sessions for the students on how to do the self-regulation techniques before taking tests.

Although only Native American students are reported here, students from all ethnicities and races reported benefits. Based on test results for all students, it was determined that practicing the HeartMath techniques increased test scores by an average of 17 points. Following implementation of the HeartMath self-regulation tools in 2003, the average attrition rate for Native American nursing students between 2003 and 2008 was 37%, compared to the 57% attrition rate in the 1997-2001 period. During this time, requirements for admission and graduation became more stringent and required increased testing. From the start of the program through 2006, the overall attrition rate for the school dropped from the 9% reported in 2001, varied from 3% or less. Use of HeartMath while in the school helped decreased the attrition rate by about 35% for Native American students from 2003 to 2008.

Students reported increased confidence in their test-taking abilities and reported fewer physical health issues, which they attributed to the regular practice of the self-regulation skills they learned.

In addition, the Native American nursing students using the stress-reducing practices demonstrated improved test-taking and perceived physical health and higher graduation rates than those who did not use them.[314]

Improving Learning and Math Proficiency in College Students

A major challenge in our current educational system is the significant number of students entering college who do not meet the basic minimum academic requirements for enrolling in college-level courses.

These students must take remedial courses in core academic subjects before they are ready to enter regular college classes. Personnel at the University of Cincinnati Clermont College (UCCC) have observed that as many as 92% of incoming first-year college students score below standards for college-level mathematics. As a result, students have difficulty succeeding in their required mathematics courses. UCCC and the Greater Cincinnati Tech Prep Consortium have formed a partnership to help solve this problem, with the goal of reducing the need for remediation in math. Working with this program, UCCC professors Drs. Michael Vislocky in mathematics and Ron Leslie in psychology pioneered a new approach.[315]

They integrated HeartMath's self-regulation techniques and heart-rhythm coherence feedback technology into college prep readiness programs in math. The goal of these programs was to reduce high school students' anxiety related to learning math and taking high-stakes tests and thereby improve students' learning, comprehension and retention.

"Lots of people are afraid of math. Learning to center in stressful situations can help these students perform better on tests, and it also opens up their life choices. So many people will switch majors just to avoid a specific math class."

*– Dr. Michael Vislocky
professor of mathematics
University of Cincinnati Clermont College*

The following elements were included in the training: 1) Discussion of the physiology of emotions. 2) Discussion of core values and engaging students in experiences that allowed them to practice sharing heartfelt emotions emerging from their core values. 3) Practice moving from the state of thinking about positive emotional experiences to actually experiencing those emotions. 4) Gaining self-awareness of emotional shifts. 5) Working in small groups to build a sense of community in order to become comfortable generating and sharing positive emotions. The overall goal of this instruction was to introduce students to the relationship between emotions and cognitive performance. To assess changes in math performance, each class of students took the Compass college placement test in mathematics at the beginning and end of the three-week program.

Heart-rhythm coherence feedback using the Freeze-Framer (now called emWave Pro) while simultaneously working on math problems was a key element of the program. In these sessions, students were able to observe their reactions to difficult problems, as reflected in the HRV feedback, thereby gaining more insight into their emotional responses and how to self-regulate them. They practiced self-activating coherence and using their intuition to find ways to solve math problems.

Students were extremely responsive to the program, and results over the years continuously improved as the professors discovered new ways to integrate and sustain students' use of the self-regulation techniques. In the first year the professors integrated HeartMath techniques, they found an average increase of 19% in math scores, which had increased to 24% gains by the third year in Compass test scores, compared to classes that had not integrated the self-regulation techniques. These gains were notable given the program's short duration and its primary focus on emotion-management skills rather than on formal math instruction.

In the fourth year, use of the self-regulation techniques and other HeartMath practices became fully integrated into the classroom program and produced the best results of all four years, far exceeding even the instructors' expectations. These results are shown in Figure 9.6, which compares students' scores on the Compass college placement test in Algebra before the program and seven weeks later, after learning and using the techniques and HRV coherence technology as an integrated part of their math instruction. The *results showed a significant ($p < 0.001$) improvement with an average increase of 73% in student math scores.*

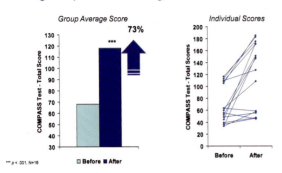

Figure 9.15. The average and individual student improvements in scores on the Compass college placement test in algebra before and seven weeks after learning and using the HeartMath self-regulation skills and HRV coherence technology as an integrated part of math instruction. Results show a statistically significant ($p < 0.001$) average increase of 73% in student scores on the college placement test.

The college prep readiness program was then expanded in following years with the intent of eliminating the need for students to have to take math remediation classes. With the support of a high school principal and math teacher, the program was integrated into an 11th grade math class at a local high school. Guided by the program instructors, 16 students and their math teacher learned the tools of the HeartMath System. The teacher guided the students in using the HeartMath techniques as an integral part of the class and homework. Four Freeze-Framer stations were set up in the classroom and students practiced using it by rotating throughout the class period.

"I'm ecstatic! High school students are excited about this program because it gives them an edge in learning math and in demonstrating their proficiency on tests. Many students who felt discouraged about their math performance now feel confident that they can succeed."
– Dr. Michael Vislocky

Drs. Vislocky and Leslie observed:

"There was a seamless integration of learning math and HeartMath infused with the curriculum in the context of the classroom setting. The teacher internalized and facilitated the HeartMath process, and actively engaged students in the learning process. The teacher got the students personally involved by giving assignments and journaling their HeartMath experiences. This provided opportunities to make continuous improvements based on feedback from students. Students were confident that their input was valued and acted upon through adjustments in the classroom."

Key factors that appear to maximize the success of the program:

> Committed teacher/facilitator.

> High expectations for student success.

> Managing emotions should take place in the context of the classroom

> Journaling or some mechanism for feedback to make continuous improvements along the way.

> Begin program at the beginning of the school year.

> Train the teacher/facilitator.

> Students must be provided with ample opportunities to apply HeartMath tools inside and outside the classroom.

> Real classroom experiences using HeartMath tools so individuals realize direct benefits.

> Integration of HeartMath in the classroom is important for promoting student-to-student interaction.

Drs. Vislocky and Leslie's work represented the first effort we knew of to integrate the HeartMath System of self-regulation tools directly into a mathematics learning environment. The findings provided strong evidence that the integration of coherence-building tools and technologies into the instruction of core academic subjects could be an effective way to enhance student learning and academic performance and to better prepare high school students for entry into higher education.

CHAPTER 10

Social Coherence: Outcome Studies in Organizations

There are obvious benefits to interacting and working with individuals who have a high level of personal coherence. When members of any work group, sports team, family or social organization get along well there is a natural tendency toward good communication, cooperation and efficiency. Social and group coherence involves the same principles as global coherence, but in this context it refers to the alignment and harmonious order in a network of relationships among individuals who share common interests and objectives, rather than the systems within the body. The principles, however, remain the same: In a coherent team, there is freedom for the individual members to do their part and thrive while maintaining cohesion and resonance within the larger group's intent and goals. Social coherence is therefore reflected as a stable, harmonious alignment of relationships that allows for the efficient flow and utilization of energy and communication required for optimal collective cohesion and action.[170]

When individuals are not well self-regulated or are acting in their own interests without regard to others, it generates social incoherence. Stressful or discordant conditions in a given group act to increase emotional stress among its members and can lead to social pathologies such as violence, abuse, inefficacy, increased errors, etc.[318] It has become increasingly clear that the leading sources of stress for adults are money issues and the social environment at work. More than 9 in 10 adults believe that stress can contribute to the development of major illnesses such as heart disease, depression and obesity, and that some types of stress can trigger heart attacks and arrhythmias. While awareness about the impact stress can have on emotional and physical health seems to be present, many working Americans continue to report symptoms of stress with 42% reporting irritability or anger, 37% fatigue, 35% a lack of interest, motivation or energy, 32% headaches and 24% upset stomachs due to stress.[317]

> **Job stress is estimated to cost American companies more than $300 billion a year in health costs, absenteeism and poor performance. In addition, consider these statistics:**
>
> - 40% of job turnover results from stress.[318]
>
> - Healthcare expenditures are nearly 50% greater for workers who report high levels of stress.[319]
>
> - Replacing an employee costs an average of 120% to 200% of the affected position's salary.[320]
>
> - An estimated 60% of all job absenteeism is caused by stress.[321]
>
> - Depression and unmanaged stress are the top two most costly risk factors in terms of medical expenditures. They increase health-care costs two to seven times more than physical risk factors such as smoking, obesity and poor exercise habits.[322]
>
> - Employees who perceive they have little control over their jobs are nearly twice as likely to develop coronary heart disease as employees with high perceived job control.[323]

It also has become apparent that social incoherence not only influences the way we feel, relate and communicate with one another, it also affects physiological processes that directly affect health. Numerous studies have found that people undergoing social and cultural changes, or living in situations characterized by social disorganization, instability, isolation or disconnectedness, are at increased risk of acquiring many types of disease.[324-328] James Lynch provides a sobering statistic on the effects of social isolation on individuals. His research in social isolation shows that loneliness produces a greater risk for heart disease than smoking, obesity, lack of exercise and excessive alcohol consumption combined.[329]

On the other hand, there is an abundance of literature showing that close relationships and social networks are highly protective. Numerous studies of diverse populations, cultures, age groups and social strata have shown that individuals who are involved in close and meaningful relationships have significantly reduced mortality, reduced susceptibility to infectious and chronic disease, improved recovery from post-myocardial infarction and improved outcomes in pregnancy and childbirth.[330-332]

There are practical steps and practices that can be taken to build and stabilize organizational coherence and resilience. There are increasing numbers of hospitals, corporations, military units, schools and athletic teams, which are actively working towards increasing their team, group or organizational coherence. We have found that collective coherence is built by first working at the individual level. As individuals become more capable of self-management, the group increases its collective coherence and can achieve its objectives more effectively.

This section contains a summary of a few examples of organizations that have provided self-regulation skills combined with heart-rhythm coherence training. Overall, the results show improved workplace communication, satisfaction, productivity, lower health-care costs, innovative problem-solving and reduced employee turnover, all of which can translate into a significant return on investment, not only financially, but also socially.

A number of hospitals that have implemented HeartMath training programs for their staff have seen increased personal, team and organizational benefits. The measures most often assessed are staff retention and employee satisfaction. For example, a study conducted at the Mayo Clinic Hospital in Phoenix, Ariz. evaluated the personal and organizational effects of the HeartMath program on reducing stress and improving the health of oncology nurses (n = 29), and clinical managers (n = 15).[333]

The compelling imperative for the project was to find a positive and effective way to address the documented high stress levels of health-care workers and explore the impact of a positive coping approach on Personal and Organizational Quality Assessment-Revised (POQA-R) scores at baseline and seven months after the training program. Personal and organizational indicators of stress decreased in the expected directions in both groups.

Figure 10.1 shows the results for the personal indicators of stress scales on the POQA-R for the oncology staff from pre-intervention to seven months post-intervention. Statistically significant differences were found for each of the personal indicators (positive outlook, gratitude, motivation, calmness, fatigue, anxiety, depression, anger management, resentfulness and stress symptoms). Figure 10. 2 shows the results for the organizational indicators of stress factors on the POQA-R for the oncology staff. Although all of the indicators trended in the expected direction, statistically significant differences were found in the indicators of goal clarity ($p < 0.01$), productivity ($p < 0.001$), communication effectiveness ($p < 0.001$) and time pressure ($p < 0.001$). Turnover on the oncology unit was 13.12% pre-intervention and 9.8% seven months post-intervention. In addition, incremental time on the oncology unit dropped from 1.19 to 0.74 during the same time interval, and employee satisfaction survey scores for the unit increased in the following areas: confidence leadership was responding to issues/concerns; confidence the organization was taking genuine interest in employees' well-being and showing a desire to continuously improve service on the unit; speaking their minds without fear; respect between physicians and allied health; and overall job satisfaction.

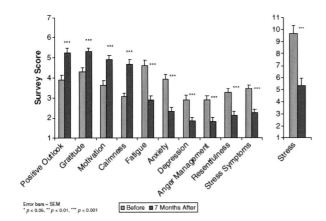

Figure 10.1 Oncology staff group, matched pairs analysis of personal indicators of stress from the Personal and Organizational Quality Assessment, at baseline and seven months post-intervention.

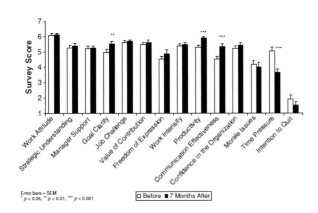

Figure 10.2 Oncology staff group, matched pairs analysis of organizational scales from the Personal and Organizational Quality Assessment, at baseline and seven months post-intervention.

Figure 10.3 depicts the results of the POQA-R on the personal indicators of stress factors for the leadership group from pre-intervention to seven months post-intervention. Statistically significant differences were found in the personal indicators of gratitude (p< 0.001), fatigue (p< 0.01), depression (p < 0.05), anger management (p<0.01), resentfulness (p<0.001) and stress symptoms (p<0.01). Figure 10.4 depicts the results of the organizational indicators of stress factors on the POQA-R for the leadership group. Statistically significant differences between baseline and seven months post-intervention were found on the indicators of manager support (p < 0.05) and value of contribution (p < 0.05).

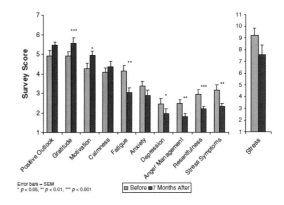

Figure 10.3 Leadership group, matched pairs analysis of personal indicators of stress from the Personal and Organizational Quality Assessment, at baseline and seven months post-intervention.

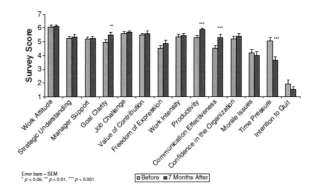

Figure 10.4 Leadership group, matched pairs analysis of organizational scales from the Personal and Organizational Quality Assessment, at baseline and seven months post-intervention.

The authors state that the findings from this study demonstrate stress and its symptoms are problematic issues for hospital and ambulatory clinic staff as evidenced by baseline measures of distress. Further, a workplace intervention was feasible and effective in promoting positive strategies for coping and enhancing well-being, personally and organizationally.

In a study conducted at the Fairfield Medical Center, a 222-bed community hospital in Lancaster, Ohio, a HeartMath workshop given in a series of six one-hour sessions, with one two-hour follow-up session was delivered to improve the well-being of hospital staff and physicians.[334] Special thought and consideration were given to being able to sustain the use of the self-regulation techniques over the long-term. As a result, strategies were developed to integrate the program into the hospital's culture.

Four staff members from a variety of disciplines were selected to be certified in the HeartMath program to gain proficiency in methodology, practices and techniques. Two Workshops each week were offered from August 2007 through December 2010. A total of 975 employees, or 48% of the staff, participated. Sustainability of the program was aided by ensuring senior leadership support, management team training, use of techniques in committee and department meetings, consulting, classes to local educators and open workshops for employee family members.

Three metrics were selected to measure the success of the program: employee satisfaction, absenteeism rates and healthcare claims cost. Significant cultural and financial return on investment was demonstrated. Employees who received the HeartMath training experienced a 2:1 savings on health-care claims, compared to employees who had not received training. Employee opinion survey results demonstrated that employees who had HeartMath training had higher overall satisfaction scores than those who had not received training. HeartMath-trained participants demonstrated a lower overall absenteeism rate, resulting in a $94,794 savings over three years.

It was concluded that the HeartMath program and sustainability practices proved to be wise decisions and continue to be valuable when initiating new concepts in a stressful, changing environment. They highlighted the fact that sustainability was the key to long-term success and a true cultural change. Continued employee training of the HeartMath techniques and continued use of the tools enriches the program planning and implementation of new initiatives at Fairfield Medical Center.

In a study conducted by the National Health Service (NHS) a publicly funded health service in the United Kingdom that provides free point-of-use services for UK residents, the HeartMath Revitalizing Care Program was provided in workshops to four departments in an NHS trust from August to October 2011.[335] Participants included staff from three clinical wards and one reception area. Over a three-month period, 97 staff members participated in the workshops. Evaluation of the project was conducted using the Personal and Organizational Quality Assessment – Revised 4 Scale (POQA-R4) as well as the pre- and post-training measures of staff turnover, absence rates and complaints.

The evaluation showed participants had demonstrated improvements in nine of the 10 personal qualities categories. These changes were statistically significant in eight areas, with fatigue and calmness showing the greatest improvements. It should be noted that the results of the pre- and post-comparison of staff turnover, sickness absence and complaints were inconclusive because of the short time frame of the study.

A study was conducted at Chesapeake Regional Medical Center (CRMC) in Virginia of 792 staff who completed pre-and post-measures with Personal and Organizational Quality Assessment (POQA) between February 2009 and December of 2010.[336] The intervention was a HeartMath workshop and training in Jean Watson's Theory of Caring (Caritas processes). As a result of incorporating Caritas processes into the HeartMath workshop, the program development team at HeartMath created a workshop called "Revitalizing Care," which integrates the Caring Theory into the HeartMath concepts and tools. There were significant improvements in positive outlook, gratitude, motivation, calmness, and anger management as well as significant reductions in fatigue, anxiety, depression, resentfulness and physical symptoms of stress. The organizational scales showed significant improvements in strategic understanding, confidence in the organization, feeling valued, freedom of expression, communication, productivity and morale issues and a decrease in intention to quit.

Caring science theory and practices have been part of Kaiser Permanente's strategic priority for the Kaiser Permanente Northern Region since 2010. Its goal is to ensure the continued spread across its medical centers of practices guided by a caring sciences framework that fosters caring-healing environments and that reinforce helping-trusting relationships between caregivers and patients. Kaiser staff were selected to become certified HeartMath trainers. There were four key elements in the trainer selection process: 1) Trainers were selected in contextual alignment with Kaiser's strategic goals. 2) Leader/RN staff relationships were important. 3) Trainers had to be committed to advancing cultures of caring science and HeartMath. 4) The chief nursing officers who would become trainers had to emphasize consistent leadership support. During a 12-month period from June 2011 to June 2012, over 400 nurses, leaders and other support staff were trained in the program. Of those 400 participants, 263 completed both the pre- and post-POQA surveys.[337] Eight of the 14 scales showed statistically significant changes in work attitude, goal clarity, communication effectiveness, time pressure, intention to quit, strategic understanding and productivity. Improvements also were noted in well-being, quality of life, impacts on patient satisfaction, safety, and reduction of absenteeism. Additionally, benefits included improved relationships between nursing staff and leaders. The trainers reported being deeply affected on professional and personal levels.

"HeartMath's programs have enabled our leaders to sustain peak performance, to manage more efficiently in a changing environment and to maintain work/life balance. For our staff, it has made the difference between required courtesy and genuine care."

—Chief Operating Officer Tom Wright, Delnor Community Hospital

---

Other examples illustrating the effects of implementing the HeartMath self-regulation program come from the Cape Fear Valley Medical System in Cape Fear, North Carolina, which reduced nurse turnover from 24% to 13%; and Delnor Community Hospital in Chicago, which experienced a similar reduction, 27% to 14%, in addition to a dramatic improvement in employee satisfaction, results that have been sustained over a 10-year period. Similarly, the Duke University Health System reduced turnover in its emergency services division from 38% to 5%.

An analysis of the combined psychometric data from 8,793 health-care workers with matched pre- and post-POQA data collected before and six weeks after completion of a HeartMath self-regulation skills training program produced many positive results. Fatigue, anxiety, depression, anger and physical stress symptoms declined greatly, while positive outlook, gratitude, motivation and calmness improved significantly (Figure 10.5).

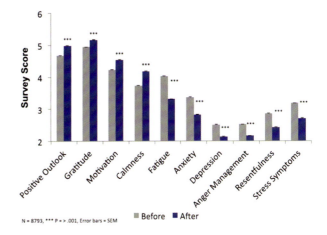

Figure 10.5 Shows matched pre- and post-data from the POQA from 8,793 health-care workers collected before and six weeks after completion of HeartMath training programs.

Studies In Law Enforcement Organizations

Several studies with police officers have found their capacity to recognize and self-regulate their responses to stressors in both work and personal contexts was significantly improved after learning the HeartMath self-regulation skills.

One study explored the nature and degree of physiological activation typically experienced by officers on the job and the effects of HeartMath's Resilience Advantage training program on a group of police officers from Santa Clara County, California.[53] Areas assessed included vitality, emotional well-being, stress coping and interpersonal skills, work performance, workplace effectiveness and climate, family relationships, and physiological recalibration following acute stressors.

Physiological measurements were obtained to determine the real-time cardiovascular effects of acutely stressful situations encountered in highly realistic simulated police calls used in police training, and to identify officers at increased risk of future health challenges. The results showed that the resilience-building training improved officers' capacity to recognize and self-regulate their responses to stressors in both work and personal contexts. Officers experienced significant reductions in stress, negative emotions and depression, compared to a control group, and increases in peacefulness and vitality, compared to the control group. (Figures 10.6 and 10.7). Improvements in family relationships, more effective communication and cooperation within work teams, and enhanced work performance also were noted.

Heart-rate and blood-pressure measurements taken during simulated police-call scenarios showed that acutely stressful circumstances typically encountered on the job resulted in a tremendous degree of physiological activation, from which it took a considerable amount of time to recover (Figure 10.8). Autonomic nervous system assessments based on heart rate variability analysis of 24-hour ECG recordings revealed that 11% of the officers were at higher risk, which was more than twice the percentage typically found in the general population.

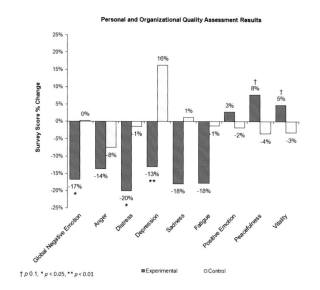

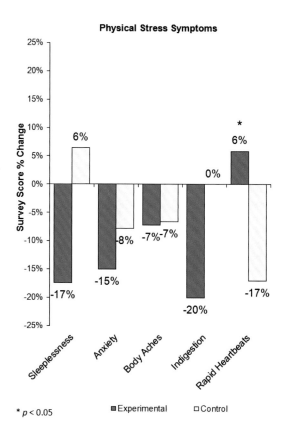

Figure 10.6. **Improvements in stress and emotional well-being.** Compares the differences between the average pre- and post-training scores for each variable as measured by the Personal and Organizational Quality Assessment. Compared to the control group, participants trained in the Resilience Advantage program exhibited significant reductions in distress, depression and global negative emotion, and increases in peacefulness and vitality. The global negative emotion score is the overall average of the individual scores for the anger, distress, depression and sadness constructs. Note that the control group experienced a marked rise in depression over the same time period. †$p < .1$, *$p < .05$, **$p < .01$.

Figure 10.7 **Changes in physical stress symptoms.** Shows the changes in five physical symptoms of stress for all participants at the start of the study and 16 weeks later (four weeks after completion of the training).

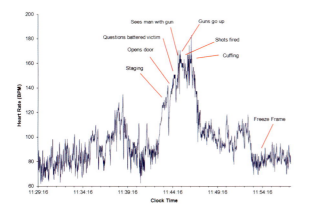

Figure 10.8 This graph provides a typical example of the ability of an officer from the experimental group to shift and reset after a domestic violence scenario. Note that when the scenario ends, the officer's heart rate initially falls, but remains elevated in a range above the normal baseline When this officer used the Freeze Frame Technique, there was an immediate, further reduction in the heart rate back to baseline. In pre-training scenarios it took on average, 1 hour and 5 minutes before returning to baseline values.

> **Key Benefits of the HeartMath Training for Police Officers***
>
> - *Increased awareness and self-management of stress reactions.*
> - *Greater confidence, balance and clarity under acute stress.*
> - *Quicker physiological and psychological recalibration following acute stress.*
> - *Improved work performance.*
> - *Reduced competition, improved communication and greater cooperation within work teams.*
> - *Reduced distress, anger, sadness and fatigue.*
> - *Reduced sleeplessness and physical stress symptoms.*
> - *Increased peacefulness and vitality.*
> - *Improved listening and relationships with family.*
>
> *Compiled from the results of psychological and performance assessments, and a semi-structured interview conducted post-training.

A study was conducted with 10 male and four female police officers and two dispatchers with the San Diego Police Department. It included self-regulation skills training comprising an introductory two-hour training session, and telephone mentoring sessions spread over a four-week period by experienced HeartMath mentors. In addition, the officers were issued an iPad app called Stress Resilience Training System (SRTS), which includes training modules on stress and its effects, HRV coherence biofeedback, a series of HeartMath self-regulation techniques and HRV-controlled games.

Outcome measures were the Personal and Organizational Quality Assessment (POQA) survey, the mentors' reports of their observations and records of participants' comments from the mentoring sessions. The POQA results were overwhelmingly positive: All four main scales showed improvement, including emotional vitality, up 25% ($p=0.05$), and physical stress, up 24% ($p=0.01$). Eight of the nine subscales showed improvement, with the stress subscale improving approximately 40% ($p=0.06$). Participant responses also were uniformly positive and enthusiastic. Individual participants praised the program and related improvements in both on-the-job performance and personal and familial situations.[193]

Fighter Pilot Study

A study of 41 fighter pilots engaging in flight simulator tasks to better understand pilots' workload and visual attention in the cockpit while conducting a simulated air-to-air tactic operations found a significant correlation between higher levels of performance and heart-rhythm coherence as well as lower levels of frustration. The study also found that the objective measurement of workload and attention distribution, as reflected in HRV coherence and eye-tracking, provided more reliable indicators than self-report approaches and that HRV coherence scores of expert pilots and novice pilots were significantly different.[191]

HeartMath in a Prison Environment

A study conducted by Lori Bosteder and Sara Hargrave in the Coffee Creek Women's Correctional Facility, in Wilsonville, Ore. sought to determine whether emotional intelligence training would help female inmates better manage the emotional difficulties of living and learning within a prison environment.[338] Many of the women who enter prison have difficulty in emotion self-management and positive social engagement. They often come from dysfunctional or abusive backgrounds, and many have become accustomed to abusing drugs and/or alcohol as a means of coping with distressing emotions. The highly charged emotional environment of a women's prison is augmented by the inherent struggles of prison life, including the loss of most freedoms and nearly all privacy. This emotional chaos frequently impairs inmates' ability to learn and acquire new skills in prison work-based education programs, and may also deter their successful transition back into society.

The study participants were 17 women in minimum security. These women were all students of the prison's 18-month work-based education program in computer technology. Over a 10-week period, the women participated in three workshops that focused on basic emotional intelligence concepts and skills in the areas of self-awareness, self-management, social awareness and relationship management. HeartMath material, practice of its tools and heart-rhythm coherence feedback were the major focus of the training. Participants read and completed self-reflective exercises in HeartMath's *Transforming Stress, Transforming Anxiety* and *Transforming Anger* books. Practice of the Neutral and Heart Lock-In self-regulation techniques was integrated into the daily classes, as was heart-rhythm coherence feedback training with handheld emWave devices. Each woman also received individualized coaching and training in HeartMath emotion-management tools and use of the heart-rhythm feedback technology. The outcome measures used in the study were the Emotional Intelligence Appraisal-ME, the Brief Symptom Inventory (BSI) self-report survey, discipline reports for the participants that were recorded by correctional officers four months before the training and four months after. In addition, observation notes over the 10-week study period were recorded by the computer technology class instructor, who noted student comments and any changes in their behaviors and interactions as they practiced the emotional awareness and self-management skills. Also, each participant's extensive evaluation at the end of the study provided both quantitative and qualitative data on her experiences in the program.

Analysis of the pre- and post-data showed significant reductions in symptoms of emotional distress, obsessive-compulsive patterns, depression and anxiety on the BSI, and significant increases in self-awareness, self-management, social awareness and relationship management on the Emotional Intelligence Appraisal scales. The discipline reports showed a significant decrease in discipline problems outside the classroom.

Qualitative data also supported the program's effectiveness. Most participants reported in their evaluations significant benefits to themselves and their social relationships. A few reported an initial increase in depression and anxiety as a result of *feeling* for the first time in years, but they improved with support from the instructor. The computer technology course instructor observed major improvement in the learning environment and growth in her students' ability to self-regulate and get along with others. Two students who were paroled after the study contacted the trainers and reported that they were using the HeartMath skills to manage the struggles they were facing on the outside and that they were creating better outcomes for themselves than they would have in the past.

The researchers observed that crucial factors in the program's success were a supportive group and the consistent support and encouragement provided by the course instructor as students experienced the initial disorientation created by the psychological changes resulting from their emotion-management practice. Because the group met daily during the week, a classroom setting worked well to provide this close attention and support. The program led to important psychosocial improvements in the participants over a relatively brief period of time and appeared to significantly enhance inmates' capacity to learn within the prison setting and to eventually successfully reintegrate into society.

CHAPTER 11

Global Coherence Research: Human-Earth Interconnectivity

Every cell in our body is bathed in an external and internal environment of fluctuating invisible magnetic forces that can affect virtually every cell and circuit in biological systems. Therefore, it should not be surprising that numerous physiological rhythms in humans and global collective behaviors are not only synchronized with solar and geomagnetic activity, but disruptions in these fields can create adverse effects on human health and behavior.

The most likely mechanism for explaining how solar and geomagnetic influences affect human health and behavior are a coupling between the human nervous system and resonating geomagnetic frequencies, Schumann resonances, which occur in the earth-ionosphere resonant cavity, Alfven waves and other very low-frequency resonances. It is well established that these resonant frequencies directly overlap with those of the human brain, cardiovascular and autonomic nervous systems.

In order to conduct research on the potential interactions between human health and behavior, a global network of 12 ultrasensitive magnetic field detectors, specifically designed to measure the earth's magnetic resonances, are being installed strategically around the planet. An important goal of the project is to motivate as many people as possible to work together in a more coherent and collaborative manor to elevate collective human consciousness. If we are persuaded that not only external fields of solar and cosmic origins, but also human consciousness and emotion can affect the mental and emotional states of others' consciousness, it broadens our view of what interconnectedness means and how it can be intentionally utilized to shape the future of the world we live in. It implies that our attitudes, emotions and intentions matter and that coherent, cooperative intent can have an important influence on global events and the quality of life on Earth.

THE GLOBAL COHERENCE INITIATIVE

The Global Coherence Initiative (GCI) was launched by HeartMath Institute in 2008. It is a science-based, co-creative initiative that has the goal to unite millions of people globally in heart-focused care and intention.

GCI employs several strategies to help increase personal, social and global coherence. An internet-based network connects people globally who want to participate in creating a shift in global consciousness. In 2015, over 160,000 people in 154 countries were involved in the initiative. Members of GCI, known as GCI ambassadors, receive regular updates informing them where to direct their energetic contributions of heart-focused care and intention. GCI also helps to educate the global community by providing tools and technologies for increasing individual, social and global coherence.

> **The following GCI hypotheses guide the ongoing collaborative research:**
>
> 1. Human and animal health, cognitive functions, emotions and behavior are affected by planetary magnetic and energetic fields.
>
> 2. The earth's magnetic fields are carriers of biologically relevant information that connects all living systems.
>
> 3. Each individual affects the global information field.
>
> 4. Large numbers of people creating heart-centered states of care, love and compassion will generate a more coherent field environment that can benefit others and help offset the current planetarywide discord and incoherence.

Embedded within the above hypotheses is a related hypothesis that human emotions and consciousness interact with and encode information in the geomagnetic field. Thereby, information is communicated nonlocally between people at a subconscious level, which, in effect, links all living systems and influences collective consciousness. Thus, a feedback loop exists between all human beings and the earth's energetic systems. It is further proposed that when coherently aligned individuals are intentionally creating physiologically coherent waves, they more effectively resonate with and encode information in the planetary magnetic fields. These magnetic fields act as carrier waves, thereby positively influencing all living systems contained within the field environment and the collective consciousness.[339]

Global Coherence Monitoring System

The Global Coherence Monitoring System (GCMS) gathers scientific data on Earth's electromagnetic fields, with state-of-the-art magnetometers located on suitable sites around the world. GCI uses the GCMS to measure and explore fluctuations and resonances in the earth's magnetic fields and in the earth-ionosphere resonant cavity in order to conduct research on the mechanisms of how the earth's fields affect human mental and emotional processes, health and collective behavior. In addition, we hope to investigate whether changes in the earth's magnetic fields occur before natural catastrophes such as earthquakes, volcanic eruptions and human events such as social unrest and terrorist attacks.

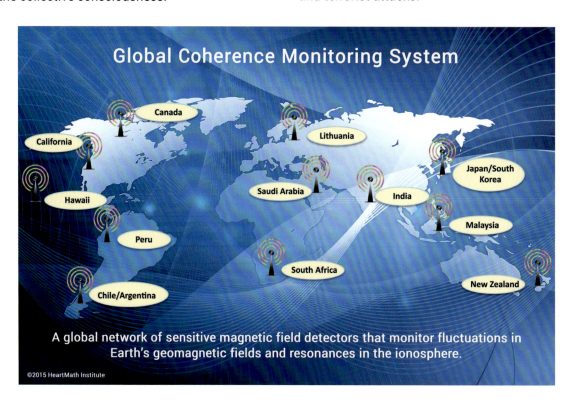

Figure 11.1. Existing and proposed locations for the global network of monitoring sites as of fall 2015. These sites are specifically designed to measure the magnetic resonances in the earth/ionosphere cavity, resonances that are generated by the vibrations of Earth's geomagnetic field lines, and ultra-low frequencies that occur in the earth's magnetic field.

This system is the first global network of GPS time-stamped detectors designed to continuously measure magnetic signals that occur in the same range as human physiological frequencies such as the brain and cardiovascular systems. A total of 12 magnetometers are planned to complete the global network. Each GCMS site includes ultrasensitive magnetic field detectors specifically designed to measure the magnetic resonances in the earth/ionosphere cavity, resonances that are generated by the vibrations of the earth's geomagnetic field lines and ultra-low frequencies that occur in the earth's magnetic fields, all of which have been shown to affect human health, mental and emotional processes and behaviors.

Each monitoring site detects the local alternating magnetic field strengths over a relatively wide frequency range (0.01-300 hertz) while maintaining a flat-frequency response. There are several networks of ground-based fluxgate magnetometers around the world, along with several space weather satellites, which measure the strength of the earth's magnetic fields and geomagnetic disturbances (Kp).

The GCI monitoring system helps us better understand how people and animals are affected by the rhythms and resonant frequencies in Earth's magnetic field as well as enabling us and other researchers to better understand the interconnections between solar and other external forces on the planetary magnetic field environment. Figure 11.2 shows a photo of the monitoring site located in Boulder Creek, California. At the time of this writing, six sites were operational. They are located at the HeartMath Research Center in Northern California, the eastern province of Saudi Arabia, Lithuania, Canada, North Island of New Zealand, and the east coast of South Africa.

Figure 11.2. The monitoring site at the HeartMath Research Center, located in Boulder Creek, California.

The data acquisition infrastructure captures, stamps with time and global positioning data, and transmits the data to a common server. In addition, each site has a random number generator (RNG) that is part of the Global Consciousness Project (GCP) network. The monitoring system tracks changes in geomagnetic activity due to solar storms, changes in solar wind speed, disruption of Schumann resonances (SR) and, potentially, the signatures of major global events that have a strong emotional component. A growing body of data also suggests that changes occur in ionospheric activity before earthquake activity.[340, 341] We make our data freely available to other research groups who may wish to explore how it may be utilized to predict earthquakes and other events. Thus, the network will provide a significant research tool to further understand how solar and geomagnetic disturbances and rhythms affect human health, emotions, behaviors and consciousness and vice versa.

Earth's Energetic Systems and Human Health and Behavior

Every cell in our bodies is bathed in an external and internal environment of fluctuating invisible magnetic forces.[205] Because fluctuations in magnetic fields can affect virtually every circuit in biological systems,[5, 205, 342] human physiological rhythms and global behaviors are not only synchronized with solar and geomagnetic activity, but disruptions in these fields can create adverse effects on human health and behavior.[343-345] Research by Burch et al.[346] and Rapoport et al.[347] provide evidence that melatonin levels are reduced during increased solar and geomagnetic activity. Cancer, neurological diseases, acute heart disease and heart attacks among other diseases and accelerated aging are all related to melatonin levels that are too low. In addition, clinical measurements have identified significant changes in blood pressure, blood flow, aggregation and coagulation, cardiac arrhythmia and heart rate variability during increased geomagnetic activity events, all of which are influenced by melatonin levels.[232, 343]

EEG patterns, heart rate, blood pressure and reaction times were measured in a group of people by Doronin et al.[343] *The authors noted that the oscillations in the magnetic field data had identical periods in the monitored EEG alpha rhythm. This confirms that whole-body changes occur in conjunction with geomagnetic activity which are reflected in changing heart and brain rhythms.*

Another study by Pobachenko et al,[348] monitored the Schumann resonances (SR) and the EEG in a frequency range of 6 to 16 hertz simultaneously. During a daily cycle, individuals studied showed variations in the EEG similar to changes in the SR. Hence, the biological EEG rhythm is characteristic of the daily rhythm of the SR.[348] The lowest frequency SR is approximately 7.83 hertz, with a daily variation of about ± 0.5 hertz. The other frequencies are ~ 14, 20, 26, 33, 39 and 45 hertz. Figure 11.3 shows the frequencies of the SR, which are closely overlapping with alpha (8 to 12 hertz), beta (12 to 30 hertz) and gamma (30 to 100 hertz) brain waves.

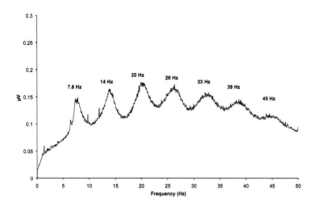

Figure 11.3. Schuman resonance data recorded from the GCI sensor site in Boulder Creek, Calif.

Because the brain is a very sensitive electromagnetic organ, changes in geomagnetic activity and SR intensities appear to alter brain-wave and neurohormone responses. Geomagnetic storms are also related to human health effects and death.[349, 350] Altered EEG rhythms have been observed by Belov et al.[351] While low-frequency magnetic oscillations (around 3 hertz) had a sedative effect in the Pobachenko et al. study, stronger oscillations of around 10 hertz stresses and stimulates people.[351]

Increased solar activity can disturb the biological rhythm of humans and exacerbate existing diseases. However, deviations are observed for some individuals, which can be caused by the individual's adaptive ability. Increased solar activity and geomagnetic activity also is correlated to a significant increase in heart attacks and incidence of death, myocardial infarction incidence,[352] a 30% to 80% increase in hospital admissions for cardiovascular disease, cardiovascular death, depression, mental disorders, psychiatric admission, suicides, homicides and traffic accidents.[344, 353-357]

Birthrates were observed in the Pobachenko et al. study to drop and mortality rates to increase during increased solar and geomagnetic activity (GMA), and migraine attacks could be triggered.[358]

Persinger and Halberg have independently shown that war and crimes were correlated to GMA.[359] Additionally, research has indicated that an increase in magnetic Pc frequencies (continuous pulsations), can affect the human cardiovascular system because Pc-1 frequencies are in a comparable range with those of the human cardiovascular system and rhythms.[360]

A study carried out in India on animals and humans also demonstrated that humans and animals can be affected by Pc frequencies.[361] The experiments showed changes in the electrophysiological, neurochemical and biochemical parameters. The subjects experienced uneasiness, confusion, restlessness and a lack of a sense of well-being when subjected to the pulsating fields. Some also complained of headaches.[361] It is important to note that of all the bodily systems studied, rhythms of the heart and brain thus far appear to be most strongly affected by changes in geomagnetic conditions.[205, 348, 349, 357, 362-367]

Historically, many cultures believed their collective behavior could be influenced by the sun and other external cycles and influences. This belief has proven to be true. On a larger societal scale, increased violence, crime rate, social unrest, revolutions and frequency of terrorist attacks have been linked to the solar cycle and the resulting disturbances in the geomagnetic field.[345, 359, 368-371] The first scientific evidence of this belief was provided by Alexander Tchijevsky, a Russian scientist who noticed that more severe battles during World War I occurred during peak sunspot periods.[371] He conducted a thorough study of global human history dating back to 1749 and compared the period to the solar cycles through the period, up to 1926. Figure 11.4, reconstructed from Tchijevsky's original data, plots the number of significant human historical events compared to the solar cycle from 1749 to 1926.[371]

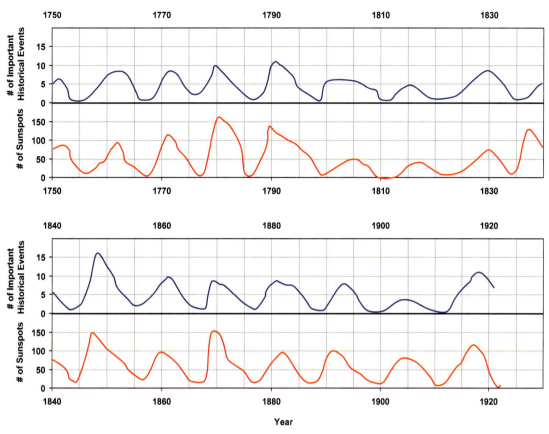

Figure 11.4. Tchijevsky's original data. The blue line plots the yearly number of important political and social events such as the start of a war, social revolutions, etc. The red line plots solar activity, as indicated by the number of sunspots from 1749 to 1922. The histories of 72 countries were compiled, and it was found that 80% of the most significant events occurred during the solar maximum, which correlates with highest periods of geomagnetic activity.

Energetic Influxes and Human Flourishing

Solar activity, in addition to being associated with social unrest, also has been related to the periods of greatest human flourishing with pronounced spurts in architecture, arts, science and positive social change.[372] We can learn from past mistakes and consciously choose new ways of navigating energy influxes to create periods of human flourishing and humanitarian advances. When outdated structures that do not serve humanity collapse, an opportunity opens for them to be replaced with more suitable and sustainable models. Such positive change can affect the political, economic, medical and educational systems, as well as relationships of individuals at work and home and in communities. At times of such heightened energy influx, we have the greatest opportunity to create positive change in our world. We can learn from past mistakes and consciously choose new ways of navigating energy influxes to create periods of human flourishing and advances.

It is well established that the earth and ionosphere generate a symphony of resonant frequencies that directly overlap with those of the human brain and cardiovascular system. The central hypothesis is that changes in these resonances can in turn influence the function of the human autonomic nervous system brain, and cardiovascular system.

Interconnectedness Study

Data and results from the Interconnectedness Study were presented in McCraty et al., 2012.[339] In 2010, 1,643 Global Coherence Initiative members from 51 countries completed a twice-weekly survey at random times six days each week over a six-month period. The survey contained six valid scales: positive affect, well-being, anxiety, confusion, fatigue and physical symptoms. The survey data were subjected to correlation analysis with a number of planetary and solar activity variables such as solar wind speed, magnetic field and plasma data, measures of energetic protons, solar flux and geomagnetic activity indices. When solar wind speed, Kp, Ap (Kp and Ap magnetic indices were designed to describe variations in the geomagnetic field) and polar cap activity increased, positive affect among the participants decreased. Well-being scores were negatively correlated with solar wind speed, Kp-index, Ap-index and polar cap magnetic activity. Thus, when solar wind speed increased and the geomagnetic field was disturbed, the levels of fatigue, anxiety and mental confusion increased. The study also resulted in some unexpected findings. For example, the solar radio flux index was positively correlated with reduced fatigue and improved positive affect, indicating there are mechanisms that improve human well-being that are not yet fully understood. Clearly, additional research needs to be conducted in order to understand the effects of the various variables and the time sequence of their effects.[339]

Examples of Magnetometer Data

Data collected by our magnetometers in different locations is providing some new insights into globally correlated activity and significant local differences.

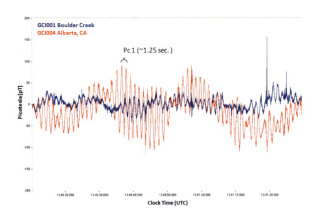

Figure 11.5. Simultaneously recorded data from Boulder Creek, Calif. and Alberta, Canada, sites.

Figure 11.5 shows an example of Pc 1 activity detected at the California and Canada sites. While the Pc 1 data in Canada displays a greater amplitude, and while most of the rhythm is synchronized, there are periods in which it is ~180 degrees out of phase. Further data processing is currently under way to examine other parameters in greater depth, such as longitudinal and latitudinal parameters, time of the day and other solar and geomagnetic parameters and their implications on human health indicators.

HRV Studies

Among physical environmental variables affecting biological processes and human health, the natural variation in the geomagnetic field in and around the earth reportedly has been involved in relation to several human cardiovascular variables. These include blood pressure,[373] heart rate (HR) and heart rate variability (HRV).[375, 376] Although there is mounting evidence for such effects, they are far from being fully understood. Several studies have found significant associations between geomagnetic storms and decreased heart rate variability (HRV), indicating a possible mechanism linking geomagnetic activity with increased incidents of coronary disease and myocardial infarction. [350, 352, 355] One study that analyzed weeklong recordings found a 25% reduction in the VLF rhythm during magnetically disturbed days, compared to quiet days. The LF rhythm also was reduced significantly, but the HF rhythms were not.[376]

In order to further investigate the potential correlations between solar and magnetic factors and HRV, we undertook a collaborative study with Dr. Abdullah A. Al Abdulgader, director of Prince Sultan Cardiac Center in Al Ahsa, Saudi Arabia, spanning a five-month period. A total of 960 24-hour HRV recordings were obtained from a group of 16 women aged 24 to 31 (mean age 31). HRV data was collected for 24 hours a day, three consecutive days each week over five months with HRV recorders between March and August of 2012. The HRV measures assessed were the interbeat-interval (IBI), SDNN, RMSSD, total power, very-low-frequency (VLF), low-frequency (LF) and high-frequency (HF) power, and the LF/HF ratio. The solar activity and magnetic variables were: solar wind speed, Kp and Ap index, PC(N), sunspot number, solar radio flux (f10.7), cosmic rays, Schumann resonance integral (area under the curve around 7.8 hertz) and the mean and standard deviation (SD) of the time-varying magnetic field data collected at GCI sites in Boulder Creek, Calif. (GCI 1) and Al Ahsa, Saudi Arabia (GCI 2). The mean and standard deviation were computed hourly. The mean field variation reflects ultra-low frequency changes and SD, which is highly correlated with total spectral power and reflects overall variance in the field. Circadian effects were removed from both environmental and HRV variables. For each of the 16 study participants, a correlation matrix was calculated between each environmental and HRV variable. Overall, the study confirms that autonomic nervous system activity as reflected by HRV measures is affected by solar and geomagnetic influences. All of the HRV measures, with the exception of IBIs, were negatively correlated with solar wind speed, and LF and HF power were negatively correlated with the magnetic field mean data collected from the Saudi Arabia site, but not the California site, suggesting that local measurements are important. Surprisingly, there were a number of positive correlations. The f10.7 index was correlated with increased HRV in all measures with the exception of the SD of the HRV and IBIs. The SD of the magnetic field variation from both the Saudi Arabia and California sites was positively correlated with RMSSD and HF power, both of which reflect parasympathetic activity, and Schumann resonance power was positively correlated with the IBIs.

Although there were a number of global correlations, at the individual level, the HRV responses varied and in some cases different individuals showed different responses to the same environmental variable.

When looking at the data from both the *Interconnectedness Study* and the HRV data, it is clear that when the earth's magnetic field was calmer or the solar radio flux was increased, the study participants felt better, were more mentally and emotionally stable and had higher levels of HRV. Conversely, when the magnetic field was disturbed, HRV was lower and participants' emotional well-being and mental clarity were adversely affected.

Figure 11.6 shows an example of healthy participants HRV-HF power plotted along with the total magnetic power spectrum from the magnetometer site in California over a 30-day period. This data is from a study of 10 participants located in northern California whose HRV was continuously monitored over a 30-day period. The magnetic field data in the plot, which is inversely correlated, has been inverted in the plot to help illustrate the visual correlation, which can be clearly seen in the graph.

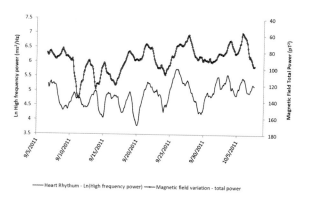

Figure 11.6. Example of one participant's high-frequency power derived from the individual's HRV and the total power of the time varying magnetic field at the California site over a 30-day period.

Interconnectivity of All Living Systems through the Earth's Magnetic Field

Magnetic Fields Carry Biologically Relevant Information

The evidence that human health and behavior are globally influenced by solar and geomagnetic activity is relatively strong and convincing. We also have shown in our laboratory that the electromagnetic field of an individual's heart can be detected by nearby animals or the nervous systems of other people.[378] (Also, see the Energetic Communication chapter in this document).

GCI hypothesizes that the earth's magnetic fields are carriers of biologically relevant information that connect all living systems. Thus, we each affect the global information field.

There is experimental evidence that human bio-emotional energy can have a subtle, but significant (scientifically measurable) nonlocal effect on people, events and organic matter.[339] It is becoming clear that a bioelectromagnetic field such as the ones radiated by the human heart and brain of one person can affect other individuals and the "global information field environment." For example, research conducted in our laboratory has confirmed the hypothesis that when an individual is in a state of heart coherence, the heart radiates a more coherent electromagnetic signal into the environment and that individual is more sensitive to detecting the information in the fields radiated by others.[378]

Of all the organs, the heart generates the largest rhythmic electromagnetic field, one that is approximately 100 times stronger than the one the brain produces. This field can be detected several feet from the body with sensitive magnetometers. This magnetic field provides a plausible mechanism for how we can "feel" or sense another person's presence and emotional state independent of body language or other factors. We also have found there is a direct relationship between the heart-rhythm patterns and the spectral information encoded in the frequency spectra of the magnetic field radiated by the heart. Thus, information about a person's emotional state is encoded in the intervals between the heartbeats, which is communicated throughout the body and into the external environment.[378]

In a study on interpersonal effects of nonverbal compassionate communication, measuring psychophysiological effects, Kemper and Shaltout found significant changes in the receiver's autonomic nervous system.[379] A growing body of evidence suggests that an energetic field is formed among individuals in groups through which communication among all the group members occurs simultaneously. In other words, there is an actual "group field" that connects all the members.[59]

Morris[221] studied heart coherence in a group setting: He investigated how people trained in maintaining states of heart coherence for several minutes might influence participants untrained in heart coherence. The results showed that the coherence of untrained participants was indeed promoted by participants in a coherent state.

Further support for the hypothesis that magnetic fields are carriers of biologically relevant information comes from a study conducted by Montagnier et al.[380] Montagnier discovered that epigenetic information related to DNA could be detected as electromagnetic signals in a highly diluted solution and that this information could be transferred to and imprinted in pure water that had never been exposed to DNA. Furthermore, this information can instruct the re-creation of DNA when the appropriate basic constituents of DNA are present and extremely low electromagnetic frequency fields of 7.8 hertz are present. They also showed that the presence of the magnetic field was needed for the information transfer to occur. The authors also state, a very low electromagnetic frequency field that transfers DNA information could come from natural sources such as the Schuman resonances (SR), which start at 7.83 hertz.

Michael Persinger, a well-known neuroscientist, also has conducted numerous studies examining the effects of magnetic fields with the same magnitude as the geomagnetic field on brain functions and information transfer.[349, 364] Not only has he shown that applying external fields similar to the SRs can induce altered states of consciousness, but he also has suggested in a detailed theory that the space occupied by the geomagnetic field can store information related to brain activity and that this information can be accessed by the human brain.[381] Persinger also suggests that the earth's magnetic field can act as a carrier of information between individuals and this information, rather than the signal intensity, is important for interaction with neural networks.[382] The above findings clearly support part of our hypothesis: *The earth's magnetic fields are carriers of biologically relevant information.* We are further suggesting that because humans have brain and heart frequencies overlapping the earth's magnetic field, not only are we receivers of biologically relevant information, but these frequencies also can couple information to the earth's magnetic fields and thus feed information into the global field environment.

Interconnection between the Human Energy Field, Collective Human Emotions and the Planetary Energy Field

Our fourth hypothesis states: *Large numbers of people creating heart-centered states of care, love, and compassion will generate a more coherent field environment that can benefit others and help offset the current planetary-wide discord and incoherence.*

There also is a substantial body of evidence indicating interactions between human emotions and a global field when large numbers of people have similar emotional responses to events or organized global peace meditations.[383-385] For example, quantum physicist John Hagelin, has conducted research on *the power of the collective* and concluded: "Since meditation provides an effective, scientifically proven way to dissolve individual stress and if society is composed of individuals, then it seems like common sense to use meditation to similarly diffuse societal stress."[384]

A study conducted in 1993 in Washington, DC, showed a 25% drop in the crime rate when 2,500 meditators mediated over specific periods of time,[385] which means that a relatively small group of a few thousand were able to influence a much larger group. The question was then posed that if crime rates could be decreased, could a group of meditators also influence social conflicts and wars? A similar experiment was done during the peak of the Israel-Lebanon war in the 1980s. Drs. Charles Alexander and John Davies at Harvard University organized groups of experienced meditators in Jerusalem, Yugoslavia and the United States to meditate and focus attention on the area at various intervals over a 27-month period. After controlling statistically for weather changes, Lebanese, Muslim, Christian and Jewish holidays, police activity, fluctuation in group sizes and other variant influences during the course of the study, researchers calculated the levels of violence in Lebanon decreased 40% to 80% each time a meditating group was in place, with the largest reductions occurring when the number of meditators was largest. During these periods, the average number of people killed during the war per day dropped from 12 to three, a decrease of more than 70%. War-related injuries declined by 68%. *Intensity level of conflict*, another of the study's measures, decreased by 48%.[383, 386]

Interconnection between Collective Human Emotions, Random Number Generators and the Geomagnetic Field

Former Princeton University Professor Roger Nelson, chief scientist of the Global Consciousness Project (GCP), provided further evidence of an interconnection between collective human emotionality and global events. GCP maintains a worldwide network of random number generators, (RNG) which produced results that suggest that human emotionality affects the randomness of these electronic devices in a globally correlated manner. Nelson said of the project: *The GCP is a long-term experiment that asks fundamental questions about human consciousness. It provides evidence for effects of synchronized collective attention – operationally defined global consciousness – on a world-spanning network of physical devices. There are multiple indicators of anomalous data structure, which are correlated specifically with moments of importance to humans. The findings suggest that some aspect of consciousness may directly create effects in the material world. This is a provocative notion, but it is the most viable of several alternative explanations.*[223]

Nelson also found clear evidence that larger events, defined by the number of people engaged and their level of "importance," produces larger effects on the global network. An interesting finding is a significant correlation between global events that elicit a high level of emotionality from a large part of the world's population and periods of nonrandom order generated by the RNGs.[387] For example, multiple independent analyses of the network during the terrorist attacks that took place in the United States on the morning of Sept. 11, 2001 correlate with a large and significant shift in the output of the global network of RNGs.[388] Although the mechanisms for how human emotions create more coherence in the randomness of this global network are not yet understood, the data clearly show that they do have such affects. Moreover, the data shows the odds-against-chance ratio is more than 1 billion to 1.[388]

When an event is characterized by deep and widespread compassion, the GCP effects are stronger,[223] which could be explained by the fact that compassion is related to interconnection and positive emotional engagement. When we experience true feelings of compassion, we tend to shift into a more coherent physiological state [5] and are thus radiating more coherent magnetic waves into the environment.[378] Compassion is an emotional state that brings people together and makes them coherent. We invest a small part of our individual being to connect with others and, as the GCP data indicate, with the global field environment. A study examining GCP data between 1998 and 2008 matched satellite-based interplanetary magnetic field (IMF) polarity with GCP-defined world events such as meditations, celebrations, natural catastrophes or violence. Study results suggested that RNG deviations may depend on a positive IMF polarity coinciding with emotionally significant conditions and/or entropy changes.[389]

The CGP has investigated a number of theoretical models that could potentially explain the global effect they are detecting with the network.

In summary, here is an excerpt of GCP's analysis:

Finally, a nonlinear dynamic field model proposes that individual minds are mutually interactive, and that the interactions are responsible for an emergent field which depends on individual consciousness but is not reducible to it. The model implies that the dynamic and interactive qualities of consciousness also involve subtle interactions with the physical world and that these are responsible for certain anomalous phenomena such as are found in the GCP experiment.[223]

We do not have magnetic data over a long enough period of time to investigate how multiple collective events associated with an outpouring of compassion or other positive collective feelings potentially may affect information that could be contained in the geomagnetic field. One of our goals, however, is to test the hypothesis *that large numbers of people in a heart-coherent state and holding a shared intention can encode physiologically patterned and relevant information that is carried by the earth's energetic and geomagnetic fields*. If living systems are indeed interconnected and communicate with each other via such biological and electromagnetic fields, it stands to reason that humans can work together in a co-creative relationship to consciously increase the coherence in the global field environment. Likewise, it also makes sense that the field environment distributes the information it may contain to all living systems within the field.

Of course, the idea that shared intentions of people in one location can influence others at a distance is not new. Such ideas have been the subject of numerous studies that have looked at the effects of prayer, meditation and groups sending intentions in various experimental contexts.[390-392]

How can we have such an influence on each other at a distance? There are no clear answers yet, but we hypothesize there is a unified field that interacts with and affects consciousness. We also suggest that individually generated coherent waves are more likely to be coupled to the larger collective field environment than waves from states of incoherence. The GCI theory of change is that as a sufficient number of individuals increase their personal coherence, it can lead to increased social coherence (families, teams, organizations), and as increasing numbers of social units (families, schools, communities, etc.) become more coherently aligned, it in turn can lead to increased global coherence, all of which is enabled and advanced through self-reinforcing feedback loops between humanity and the global field environment.

This implies that every individual contributes to the global field environment and each person's attitudes, intentions and emotional experiences count. This is empowering for many individuals who often feel overwhelmed by the current negative predictions and conflicts on the planet. It helps them realize that their actions and intentions can make a difference and that by increasing their own coherence, they can become *coherence builders* and make a contribution that can help accelerate the shift that many now perceive to be occurring.

The personal benefits of better emotion self-regulation, enhanced well-being, more self-responsibility, better health and improved relationships people experience are powerful motivators that reinforce the process for the individual. As more and more individuals become increasingly self-regulated and grow in conscious awareness, the increased individual coherence in turn increases social coherence, which is reflected in increased cooperation and effective co-creative initiatives for the benefit of society and the planet. From our perspective, a shift in consciousness is necessary to achieve new levels of cooperation and collaboration in the kind of innovative problem-solving and intuitive discernment required for addressing our social, environmental and economic problems. In time, increasing global coherence will be indicated by more and more communities, states and countries adopting a more coherent planetary view.

CONCLUSIONS

An ongoing goal of GCI is to further the study of interconnectedness between humanity and the earth's energetic systems. GCI conducts research on the mechanisms of how the earth's fields affect human mental and emotional processes, health outcomes and collective human behavior and explore how collective human emotions and intentions may be carried by the earth's electromagnetic and energetic fields. Toward these goals, as previously explained, our global network of ultrasensitive magnetic field detectors, specifically designed to measure the magnetic resonances in the earth/ionosphere cavity and resonances and Earth's geomagnetic field line resonances are being installed at strategic locations around the globe. We are hopeful our efforts will promote and contribute to a deeper understanding of the mechanisms by which human health and behaviors are modulated by the earth's geomagnetic fields and further clarify which aspects of the field environment mediate the varied and specific effects.

Data from the *Interconnectedness Study* and HRV studies are yielding promising results and add to the body of evidence that humans are affected by planetary energetic fields. GCI hypothesizes that human emotions and consciousness interact with and encode information in planetary energetic fields, including the geomagnetic field, thereby communicating information nonlocally between people at a subconscious level, which, in effect, links all living systems and gives rise to a form of collective consciousness. Thus, a feedback loop exists among all human beings and the earth's energetic systems.

The essence of the hypothesis is that when enough individuals and social groups increase their coherence and utilize it to intentionally create a more coherent standing reference wave in the global field, it will help to lift global consciousness. This can be achieved when an increasing ratio of people move toward more balanced and self-regulated emotions and responses. This in turn can help promote cooperation and collaboration in innovative problem-solving and intuitive discernment for addressing society's significant social, environmental and economic problems. In time, as more individuals stabilize the global field and families, workplaces and communities, etc., achieve increased social coherence, global coherence will increase. This will be indicated by countries adopting a more coherent planetary view that will lead them to address social and economic oppression, warfare, cultural intolerance, crime and disregard for the environment in more meaningful and successful ways.

BIBLIOGRAPHY

1. Gahery, Y. and D. Vigier, *Inhibitory effects in the cuneate nucleus produced by vago-aortic afferent fibers.* Brain Research, 1974. **75**: p. 241-246.

2. Wölk, C. and M. Velden, *Detection variability within the cardiac cycle: Toward a revision of the 'baroreceptor hypothesis'.* Journal of Psychophysiology, 1987. **1**: p. 61-65.

3. Wölk, C. and M. Velden, *Revision of the baroreceptor hypothesis on the basis of the new cardiac cycle effect*, in *Psychobiology: Issues and Applications*, N.W. Bond and D.A.T. Siddle, Editors. 1989, Elsevier Science Publishers B.V.: North-Holland. p. 371-379.

4. Lane, R.D., et al., *Activity in medial prefrontal cortex correlates with vagal component of heart rate variability during emotion.* Brain and Cognition, 2001. **47**: p. 97-100.

5. McCraty, R., Atkinson, M., Tomasino, D., & Bradley, R. T, *The coherent heart: Heart-brain interactions, psychophysiological coherence, and the emergence of system-wide order.* Integral Review, 2009. **5**(2): p. 10-115.

6. McCraty, R., M. Atkinson, and R.T. Bradley, *Electrophysiological evidence of intuition: Part 2. A system-wide process?* J Altern Complement Med, 2004. **10**(2): p. 325-36.

7. Svensson, T.H. and P. Thoren, *Brain noradrenergic neurons in the locus coeruleus: Inhibition by blood volume load through vagal afferents.* Brain Research, 1979. **172**(1): p. 174-178.

8. Schandry, R. and P. Montoya, *Event-related brain potentials and the processing of cardiac activity.* Biological Psychology, 1996. **42**: p. 75-85.

9. Montoya, P., R. Schandry, and A. Muller, *Heartbeat evoked potentials (HEP): Topography and influence of cardiac awareness and focus of attention.* Electroencephalography and Clinical Neurophysiology, 1993. **88**: p. 163-172.

10. Zhang, J.X., R.M. Harper, and R.C. Frysinger, *Respiratory modulation of neuronal discharge in the central nucleus of the amygdala during sleep and waking states.* Experimental Neurology, 1986. **91**: p. 193-207.

11. Armour, J.A., *Anatomy and function of the intrathoracic neurons regulating the mammalian heart*, in *Reflex Control of the Circulation*, I.H. Zucker and J.P. Gilmore, Editors. 1991, CRC Press: Boca Raton. p. 1-37.

12. Armour, J.A., *Potential clinical relevance of the 'little brain' on the mammalian heart.* Exp Physiol, 2008. **93**(2): p. 165-76.

13. Armour, J.A., *Neurocardiology--Anatomical and functional principles* 2003, Boulder Creek, CA: HeartMath Research Center, HeartMath Institute, Publication No. 03-011.

14. Armour, J.A. and J.L. Ardell, eds. *Neurocardiology*. 1994, Oxford University Press: New York.

15. Cameron, O.G., *Visceral Sensory Neuroscience: Interception* 2002, New York: Oxford University Press.

16. Kukanova, B. and B. Mravec, *Complex intracardiac nervous system.* Bratisl Lek Listy, 2006. **107**(3): p. 45-51.

17. Armour, J.A., *Peripheral autonomic neuronal interactions in cardiac regulation*, in *Neurocardiology*, J.A. Armour and J.L. Ardell, Editors. 1994, Oxford University Press: New York. p. 219-244.

18. Cantin, M. and J. Genest, *The heart as an endocrine gland.* Pharmacol Res Commun, 1988. **20 Suppl 3**: p. 1-22.

19. Strohle, A., et al., *Atrial natriuretic hormone decreases endocrine response to a combined dexamethasone-corticotropin-releasing hormone test.* Biol Psychiatry, 1998. **43**(5): p. 371-5.

20. Butler, G.C., B.L. Senn, and J.S. Floras, *Influence of atrial natriuretic factor on heart rate variability in normal men.* Am J Physiol, 1994. **267**(2 Pt 2): p. H500-5.

21. Vollmar, A.M., et al., *A possible linkage of atrial natriuretic peptide to the immune system.* Am J Hypertens, 1990. **3**(5 Pt 1): p. 408-11.

22. Telegdy, G., *The action of ANP, BNP and related peptides on motivated behavior in rats.* Reviews in the Neurosciences, 1994. **5**(4): p. 309-315.

23. Huang, M., et al., *Identification of novel catecholamine-containing cells not associated with sympathetic neurons in cardiac muscle.* Circulation, 1995. **92**(8(Suppl)): p. I-59.

24. Gutkowska, J., et al., *Oxytocin is a cardiovascular hormone.* Brazilian Journal of Medical and Biological Research, 2000. **33**: p. 625-633.

25. Hilton, J., *On the Influence of Mechanical and Physiological Rest* 1863, london: Bell and Daldy.

26. Shapiro, A.P., *Hypertension and Stress: A Unified Concept* 1996, Mahwah, NJ: Lawrence Erlbaum Associates.

27. Fauvel, J.P., et al., *Mental stress-induced increase in blood pressure is not related to baroreflex sensitivity in middle-aged healthy men.* Hypertension, 2000. **35**(4): p. 887-91.

28. Freeman, L.J., et al., *Psychological stress and silent myocardial ischemia.* Am Heart J, 1987. **114**(3): p. 477-82.

29. Lecomte, D., P. Fornes, and G. Nicolas, *Stressful events as a trigger of sudden death: a study of 43 medico- legal autopsy cases [see comments].* Forensic Sci Int, 1996. **79**(1): p. 1-10.

30. Aboa-Eboule, C., et al., *Job strain and risk of acute recurrent coronary heart disease events.* Jama, 2007. **298**(14): p. 1652-60.

31. Henry, J.P., *Mechanisms by which stress can lead to coronary heart disease.* Postgrad Med J, 1986. **62**(729): p. 687-93.

32. Cas, L.D., et al., *[Stress and ischemic heart disease].* Cardiologia, 1993. **38**(12 Suppl 1): p. 415-25.

33. Brunckhorst, C.B., et al., *[Stress, depression and cardiac arrhythmias].* Ther Umsch, 2003. **60**(11): p. 673-81.

34. Kageyama, T., et al., *Self-reported sleep quality, job stress, and daytime autonomic activities assessed in terms of short-term heart rate variability among male white-collar workers.* Ind Health, 1998. **36**(3): p. 263-72.

35. Chandola, T., E. Brunner, and M. Marmot, *Chronic stress at work and the metabolic syndrome: prospective study.* Bmj, 2006. **332**(7540): p. 521-5.

36. Griffith, L.S., B.J. Field, and P.J. Lustman, *Life stress and social support in diabetes: association with glycemic control.* Int J Psychiatry Med, 1990. **20**(4): p. 365-72.

37. Delamater, A.M., et al., *Stress and coping in relation to metabolic control of adolescents with type I diabetes.* Journal of Developmental Behavioral Pediatrics, 1987. **8**: p. 136-140.

38. Goldstein, D.S., *Stress, allostatic load, catecholamines, and other neurotransmitters in neurodegenerative diseases.* Endocr Regul, 2011. **45**(2): p. 91-8.

39. Frese, M., *Stress at work and psychosomatic complaints: a causal interpretation.* Journal of Applied Psychology, 1985. **70**(2): p. 314.

40. Gaines, J. and J. Jermier, *Emotional exhaustion in a high stress organization.* Academy of Management Journal, 1983. **26**(4): p. 567-586.

41. Fowers, B., *Perceived control, illness status, stress and adjustment to cardiac illness.* Journal of Psychology, 1994. **128**(5): p. 567-579.

42. Brotman, D.J., S.H. Golden, and I.S. Wittstein, *The cardiovascular toll of stress.* Lancet, 2007. **370**(9592): p. 1089-100.

43. Marchand, A. and P. Durand, *Psychological distress, depression, and burnout: similar contribution of the job demand-control and job demand-control-support models?* J Occup Environ Med, 2011. **53**(2): p. 185-9.

44. Fredrickson, B.L., *Positive emotions*, in *Handbook of Positive Psychology*, C.R. Snyder and S.J. Lopez, Editors. 2002, Oxford University Press: New York. p. 120-134.

45. Isen, A.M., *Positive affect*, in *Handbook of Cognition and Emotion*, T. Dalgleish and M. Power, Editors. 1999, John Wiley & Sons: New York. p. 522-539.

46. Wichers, M.C., et al., *Evidence that moment-to-moment variation in positive emotions buffer genetic risk for depression: a momentary assessment twin study.* Acta Psychiatr Scand, 2007. **115**(6): p. 451-7.

47. Fredrickson, B.L., *The role of positive emotions in positive psychology. The broaden-and-build theory of positive emotions.* American Psychologist, 2001. **56**(3): p. 218-226.

48. Fredrickson, B.L. and T. Joiner, *Positive emotions trigger upward spirals toward emotional well-being.* Psychological Science, 2002. **13**(2): p. 172-175.

49. Fredrickson, B.L., et al., *What good are positive emotions in crises? A prospective study of resilience and emotions following the terrorist attacks on the United States on September 11th, 2001.* Journal of Personality and Social Psychology, 2003. **84**(2): p. 365-376.

50. McCraty, R. and D. Tomasino, *Emotional stress, positive emotions, and psychophysiological coherence*, in *Stress in Health and Disease*, B.B. Arnetz and R. Ekman, Editors. 2006, Wiley-VCH: Weinheim, Germany. p. 342-365.

51. McCraty, R., et al., *The effects of emotions on short-term power spectrum analysis of heart rate variability.* Am J Cardiol, 1995. **76**(14): p. 1089-93.

52. Rein, G., M. Atkinson, and R. McCraty, *The physiological and psychological effects of compassion and anger.* Journal of Advancement in Medicine, 1995. **8**(2): p. 87-105.

53. McCraty, R. and M. Atkinson, *Resilence Training Program Reduces Physiological and Psychological Stress in Police Officers.* Global Advances in Health and Medicne, 2012. **1**(5): p. 44-66.

54. Luthar, S.S., D. Cicchetti, and B. Becker, *The construct of resilience: a critical evaluation and guidelines for future work.* Child Dev, 2000. **71**(3): p. 543-62.

55. Lieberman, M.D., *Social cognitive neuroscience: A review of core processes*, in *Annual Review of Psychology* 2007, Annual Reviews: Palo Alto. p. 259-289.

56. Baumeister, R.F., et al., *Self-regulation and personality: how interventions increase regulatory success, and how depletion moderates the effects of traits on behavior.* J Pers, 2006. **74**(6): p. 1773-801.

57. Antonovksy, A., *Unraveling The Mystery of Health: How People Manage Stress and Stay Well.* San Francisco: Jossey-Bass, 1987. Cited in: Tresolini, CP and the Pew-Fetzer Task Force. "Health Professions Education and Relationship-Centered Care" San Francisco: Pew Health Professions Commission and the Fetzer Institute, 1994 p. 15.1987.

58. McCraty, R. and M. Zayas, *Cardiac coherence, self-regulation, autonomic stability, and psychosocial well-being.* Frontiers in Psychology, 2014. **5**(September): p. 1-13.

59. McCraty, R., Childre, D, *Coherence: Bridging Personal, Social and Global Health.* Alternative Therapies in Health and Medicine, 2010. **16**(4): p. 10-24.

60. Nerurkar, A., et al., *When physicians counsel about stress: results of a national study.* JAMA Intern Med, 2013. **173**(1): p. 76-7.

61. Avey, H., et al., *Health care providers' training, perceptions, and practices regarding stress and health outcomes.* J Natl Med Assoc, 2003. **95**(9): p. 833, 836-45.

62. Cummings, N.A. and G.R. VandenBos, *The twenty years Kaiser-Permanente experience with psychotherapy and medical utilization: implications for national health policy and national health insurance.* Health Policy Q, 1981. **1**(2): p. 159-75.

63. Grossarth-Maticek, R. and H.J. Eysenck, *Self-regulation and mortality from cancer, coronary heart disease and other causes:*

A prospective study. Personality and Individual Differences, 1995. **19**(6): p. 781-795.

64. Pressman, S.D., M.W. Gallagher, and S.J. Lopez, *Is the emotion-health connection a "first-world problem"?* Psychol Sci, 2013. **24**(4): p. 544-9.

65. Mittleman, M.A., et al., *Triggering of acute myocardial infarction onset by episodes of anger. Determinants of Myocardial Infarction Onset Study Investigators.* Circulation, 1995. **92**(7): p. 1720-5.

66. Lyubomirsky, S., L. King, and E. Diener, *The benefits of frequent positive affect: does happiness lead to success?* Psychol Bull, 2005. **131**(6): p. 803-55.

67. Danner, D.D., D.A. Snowdon, and W.V. Friesen, *Positive emotions in early life and longevity: Findings from the nun study.* Journal of Personality and Social Psychology, 2001. **80**(5): p. 804-813.

68. Kawachi, I., et al., *Prospective study of phobic anxiety and risk of coronary heart disease in men.* Circulation, 1994. **89**(5): p. 1992-7.

69. Grossarth-Maticek, R. and H.J. Eysenck, *Creative novation behaviour therapy as a prophylactic treatment for cancer and coronary heart disease: Part I--Description of treatment [published erratum appears in Behav Res Ther 1993 May;31(4):437] [see comments].* Behav Res Ther, 1991. **29**(1): p. 1-16.

70. Kubzansky, L.D., et al., *Is worrying bad for your heart? A prospective study of worry and coronary heart disease in the Normative Aging Study.* Circulation, 1997. **95**(4): p. 818-824.

71. Rosenman, R.H., *The independent roles of diet and serum lipids in the 20th-century rise and decline of coronary heart disease mortality.* Integr Physiol Behav Sci, 1993. **28**(1): p. 84-98.

72. Penninx, B.W., et al., *Effects of social support and personal coping resources on mortality in older age: the Longitudinal Aging Study Amsterdam.* Am J Epidemiol, 1997. **146**(6): p. 510-9.

73. Allison, T.G., et al., *Medical and economic costs of psychologic distress in patients with coronary artery disease.* Mayo Clinic Proceedings, 1995. **70**(8): p. 734-742.

74. Eysenck, H.J., *Personality, stress and cancer: Prediction and prophylaxis.* British Journal of Medical Psychology, 1988. **61**(Pt 1): p. 57-75.

75. Thomas, S.A., et al., *Psychosocial factors and survival in the cardiac arrhythmia suppression trial (CAST): a reexamination.* Am J Crit Care, 1997. **6**(2): p. 116-126.

76. Siegman, A.W., et al., *Dimensions of anger and CHD in men and women: self-ratings versus spouse ratings.* J Behav Med, 1998. **21**(4): p. 315-36.

77. Carroll, D., et al., *Blood pressure reactions to the cold pressor test and the prediction of ischaemic heart disease: data from the Caerphilly Study.* Journal of Epidemiology and Community Health, 1998. **52**: p. 528-529.

78. Dolcos, F., A.D. Iordan, and S. Dolcos, *Neural correlates of emotion-cognition interactions: A review of evidence from brain imaging investigations.* J Cogn Psychol (Hove), 2011. **23**(6): p. 669-694.

79. Damasio, A.R., *Descartes' Error: Emotion, Reason and the Human Brain*1994, New York: G.P. Putnam's Sons.

80. Goleman, D., *Emotional Intelligence*1995, New York: Bantam Books.

81. McCraty, R. and D. Childre, *Coherence: bridging personal, social, and global health.* Altern Ther Health Med, 2010. **16**(4): p. 10-24.

82. Porges, S.W., *The polyvagal perspective.* Biol Psychol, 2007. **74**(2): p. 116-43.

83. Shaffer, F., R. McCraty, and C. Zerr, *A healthy heart is not a metronome: An integrative review of the heart's anatomy and heart rate variability.* Frontiers in Psychology, 2014. **5:1040**.

84. Singer, D.H., et al., *Low heart rate variability and sudden cardiac death.* Journal of Electrocardiology, 1988(Supplemental issue): p. S46-S55.

85. Singer, D.H., *High heart rate variability, marker of healthy longevity.* Am J Cardiol, 2010. **106**(6): p. 910.

86. Geisler, F.C., et al., *Cardiac vagal tone is associated with social engagement and self-regulation.* Biol Psychol, 2013. **93**(2): p. 279-86.

87. Reynard, A., et al., *Heart rate variability as a marker of self-regulation.* Appl Psychophysiol Biofeedback, 2011. **36**(3): p. 209-15.

88. Segerstrom, S.C. and L.S. Nes, *Heart rate variability reflects self-regulatory strength, effort, and fatigue.* Psychol Sci, 2007. **18**(3): p. 275-81.

89. Thayer, J.F., et al., *Heart rate variability, prefrontal neural function, and cognitive performance: the neurovisceral integration perspective on self-regulation, adaptation, and health.* Ann Behav Med, 2009. **37**(2): p. 141-53.

90. Camm, A.J., et al., *Heart rate variability standards of measurement, physiological interpretation, and clinical use. Task Force of the European Society of Cardiology and the North American Society of Pacing and Electrophysiology.* Circulation, 1996. **93**(5): p. 1043-1065.

91. Hon, E.H. and S.T. Lee, *Electronic evaluations of the fetal heart rate patterns preceeding fetal death: further observations.* American Journal of Obstetric Gynecology, 1965. **87**: p. 814-826.

92. Braune, H.J. and U. Geisendorfer, *Measurement of heart rate variations: influencing factors, normal values and diagnostic impact on diabetic autonomic neuropathy.* Diabetes Res Clin Pract, 1995. **29**(3): p. 179-87.

93. Vinik, A.I., et al., *Diabetic autonomic neuropathy.* Diabetes Care, 2003. **26**(5): p. 1553-79.

94. Ewing, D., I. Campbell, and B. Clarke, *Mortality in diabetic autonomic neuropathy.* Lancet, 1976. **1**: p. 601-603.

95. Wolf, M.M., et al., *Sinus arrhythmia in acute mycardial infarction.* Medical Journal of Australia, 1978. **2**: p. 52-53.

96. Umetani, K., et al., *Twenty-four hour time domain heart rate variability and heart rate: relations to age and gender over nine decades.* J Am Coll Cardiol, 1998. **31**(3): p. 593-601.

97. Dekker, J.M., et al., *Heart rate variability from short electrocardiographic recordings predicts mortality from all causes in middle-aged and elderly men. The Zutphen Study.* American Journal of Epidemiology, 1997. **145**(10): p. 899-908.

98. Tsuji, H., et al., *Reduced heart rate variability and mortality risk in an elderly cohort. The Framingham Heart Study.* Circulation, 1994. **90**(2): p. 878-883.

99. Berntson, G.G., et al., *Cardiac autonomic balance versus cardiac regulatory capacity.* Psychophysiology, 2008. **45**(4): p. 643-52.

100. Beauchaine, T., *Vagal tone, development, and Gray's motivational theory: toward an integrated model of autonomic nervous system functioning in psychopathology.* Dev Psychopathol, 2001. **13**(2): p. 183-214.

101. Geisler, F. and T. Kubiak, *Heart rate variability predicts self control in goal pursuit.* European Journal of Personality, 2009. **23**(8): p. 623-633.

102. Appelhans, B. and L. Luecken, *Heart Rate Variability as an Index of Regulated Emotional Responding.* Review of General Psychology, 2006. **10**(3): p. 229-240.

103. Geisler, F., et al., *The impact of heart rate variability on subjective well-being is mediated by emotion regulation.* Personality and Individual Differences, 2010. **49**(7): p. 723-728.

104. Smith, T.W., et al., *Matters of the variable heart: respiratory sinus arrhythmia response to marital interaction and associations with marital quality.* J Pers Soc Psychol, 2011. **100**(1): p. 103-19.

105. Nasermoaddeli, A., M. Sekine, and S. Kagamimori, *Association between sense of coherence and heart rate variability in healthy subjects.* Environ Health Prev Med, 2004. **9**(6): p. 272-4.

106. Zohar, A., R. Cloninger, and R. McCraty, *Personality and Heart Rate Variability: Exploring Pathways from Personality to Cardiac Coherence and Health.* Open Journal of Social Sciences, 2013. **1**(6): p. 32-39.

107. Ramaekers, D., et al., *Association between cardiac autonomic function and coping style in healthy subjects.* Pacing Clin Electrophysiol, 1998. **21**(8): p. 1546-52.

108. Lloyd, A., Brett, D., Wesnes, K., *Coherence Training Improves Cognitive Functions and Behavior In Children with ADHD.* Alternative Therapies in Health and Medicine, 2010. **16**(4): p. 34-42.

109. Ginsberg, J.P., Berry, M.E., Powell, D.A., *Cardiac Coherence and PTSD in Combat Veterans.* Alternative Therapies in Health and Medicine, 2010. **16**(4): p. 52-60.

110. Bradley, R.T., et al., *Emotion self-regulation, psychophysiological coherence, and test anxiety: results from an experiment using electrophysiological measures.* Appl Psychophysiol Biofeedback, 2010. **35**(4): p. 261-83.

111. Lehrer, P.M., et al., *Heart rate variability biofeedback increases baroreflex gain and peak expiratory flow.* Psychosomatic Medicine, 2003. **65**(5): p. 796-805.

112. Bedell, W., *Coherence and hearlth care cost - RCA acturial study: A cost-effectivness cohort study* Alternative Therapies in Health and Medicine, 2010. **16**(4): p. 26-31.

113. Alabdulgader, A., *Coherence: A Novel Nonpharmacological Modality for Lowering Blood Pressure in Hypertensive Patients.* Global Advances in Health and Medicne, 2012. **1**(2): p. 54-62.

114. McCraty, R., et al., *New hope for correctional officers: an innovative program for reducing stress and health risks.* Appl Psychophysiol Biofeedback, 2009. **34**(4): p. 251-72.

115. McCraty, R., M. Atkinson, and D. Tomasino, *Impact of a workplace stress reduction program on blood pressure and emotional health in hypertensive employees.* J Altern Complement Med, 2003. **9**(3): p. 355-69.

116. McCraty, R., et al., *The impact of a new emotional self-management program on stress, emotions, heart rate variability, DHEA and cortisol.* Integr Physiol Behav Sci, 1998. **33**(2): p. 151-70.

117. Oppenheimer, S. and D. Hopkins, *Suprabulbar neuronal regulation of the heart*, in *Neurocardiology*, J.A. Armour and J.L. Ardell, Editors. 1994, Oxford University Press: New York. p. 309-341.

118. Hopkins, D. and H. Ellenberger, *Cardiorespiratory neurons in the mudulla oblongata: Input and output relationhsips*, in *Neurocardiology*, J.A. Armour and J.L. Ardell, Editors. 1994, Oxford University Press: New York. p. 219-244.

119. Beaumont, E., et al., *Network interactions within the canine intrinsic cardiac nervous system: implications for reflex control of regional cardiac function.* J Physiol, 2013. **591**(Pt 18): p. 4515-33.

120. Hainsworth, R., *The control and physiological importance on heart rate*, in *Heart Rate Variability*, M. Malik and A.J. Camm, Editors. 1995, Futura Publishing COmpany, Inc.: Armonk NY. p. 3-19.

121. Palatini, P., *Elevated heart rate as a predictor of increased cardiovascular morbidity.* J Hypertens Suppl, 1999. **17**(3): p. S3-10.

122. Stampfer, H.G., *The relationship between psychiatric illness and the circadian pattern of heart rate.* Aust N Z J Psychiatry, 1998. **32**(2): p. 187-98.

123. Stampfer, H.G. and S.B. Dimmitt, *Variations in circadian heart rate in psychiatric disorders: theoretical and practical implications.* ChronoPhysiology and Therapy, 2013. **3**: p. 41–50.

124. Opthof, T., *The normal range and determinants of the intrinsic heart rate in man.* Cardiovasc Res, 2000. **45**(1): p. 177-84.

125. Umetani, K., C.L. Duda, and D.H. Singer. *Aging effects on cycle length dependence of heart rate variability.* in *Biomedical Engineering Conference, 1996.* 1996. Proceedings of the 1996 Fifteenth Southern. IEEE.

126. Electrophysiology, T.F.o.t.E.S.o.C.a.t.N.A.S.o.P.a., *Heart rate variability: Standards of measurement, physiological interpretation, and clinical use.* Circulation, 1996. **93**: p. 1043-1065.

127. Hirsch, J.A. and B. Bishop, *Respiratory sinus arrhythmia in humans: How breathing pattern modulates heart rate.* American Journal of Physiology, 1981. **241**(4): p. H620-H629.

128. Eckberg, D.L., *Human sinus arrhythmia as an index of vagal outflow.* Journal of Applied Physiology, 1983. **54**: p. 961-966.

129. Malliani, A., *Association of Heart Rate Variability components with physiological regulatory mechanisms*, in *Heart Rate Variability*, M. Malik and A.J. Camm, Editors. 1995, Futura Publishing COmpany, Inc.: Armonk NY. p. 173-188.

130. deBoer, R.W., J.M. Karemaker, and J. Strackee, *Hemodynamic fluctuations and baroreflex sensitivity in humans: a beat-to-beat model.* Am J Physiol, 1987. **253**(3 Pt 2): p. H680-9.

131. Baselli, G., et al., *Model for the assessment of heart period variability interactions of respiration influences.* Medical and Biological Engineering and Computing, 1994. **32**(2): p. 143-152.

132. Ahmed, A.K., J.B. Harness, and A.J. Mearns, *Respiratory Control of Heart Rate.* Eur J Appl Physiol, 1982. **50**: p. 95-104.

133. Tiller, W.A., R. McCraty, and M. Atkinson, *Cardiac coherence: a new, noninvasive measure of autonomic nervous system order.* Altern Ther Health Med, 1996. **2**(1): p. 52-65.

134. Brown, T.E., et al., *Important influence of respiration on human R-R interval power spectra is largely ignored.* J Appl Physiol (1985), 1993. **75**(5): p. 2310-7.

135. Malliani, A., et al., *Power spectral analysis of cardiovascular variability in patients at risk for sudden cardiac death.* J Cardiovasc Electrophysiol, 1994. **5**(3): p. 274-86.

136. Pal, G.K., et al., *Sympathovagal imbalance contributes to prehypertension status and cardiovascular risks attributed by insulin resistance, inflammation, dyslipidemia and oxidative stress in first degree relatives of type 2 diabetics.* PLoS One, 2013. **8**(11): p. e78072.

137. Pagani, M., F. Lombardi, and S. Guzzette, *Power spectral analysis of heart rate and arterial pressure variabilities as a marker of sympatho-vagal interaction in man and conscious dog.* Circulation Research, 1986. **59**: p. 178-184.

138. Axelrod, S., et al., *Spectral analysis of fluctuations in heart rate: An objective evaluation.* Nephron, 1987. **45**: p. 202-206.

139. Schmidt, H., et al., *Autonomic dysfunction predicts mortality in patients with multiple organ dysfunction syndrome of different age groups.* Crit Care Med, 2005. **33**(9): p. 1994-2002.

140. Hadase, M., et al., *Very low frequency power of heart rate variability is a powerful predictor of clinical prognosis in patients with congestive heart failure.* Circ J, 2004. **68**(4): p. 343-7.

141. Tsuji, H., et al., *Impact of reduced heart rate variability on risk for cardiac events. The Framingham Heart Study.* Circulation, 1996. **94**(11): p. 2850-5.

142. Bigger, J.T., Jr., et al., *Frequency domain measures of heart period variability and mortality after myocardial infarction.* Circulation, 1992. **85**(1): p. 164-71.

143. Shah, A.J., et al., *Posttraumatic stress disorder and impaired autonomic modulation in male twins.* Biol Psychiatry, 2013. **73**(11): p. 1103-10.

144. Lampert, R., et al., *Decreased heart rate variability is associated with higher levels of inflammation in middle-aged men.* Am Heart J, 2008. **156**(4): p. 759 e1-7.

145. Carney, R.M., et al., *Heart rate variability and markers of inflammation and coagulation in depressed patients with coronary heart disease.* J Psychosom Res, 2007. **62**(4): p. 463-7.

146. Theorell, T., et al., *Saliva testosterone and heart rate variability in the professional symphony orchestra after "public faintings" of an orchestra member.* Psychoneuroendocrinology, 2007. **32**(6): p. 660-8.

147. Kleiger, R.E., P.K. Stein, and J.T. Bigger, Jr., *Heart rate variability: measurement and clinical utility.* Ann Noninvasive Electrocardiol, 2005. **10**(1): p. 88-101.

148. Akselrod, S., et al., *Power spectrum analysis of heart rate fluctuation: a quantitative probe of beat-to-beat cardiovascular control.* Science, 1981. **213**(10): p. 220-222.

149. Cerutti, S., A.M. Bianchi, and L.T. Mainardi, *Spectral analysis of the heart rate variability signal*, in *Heart Rate Variability*, M. Malik and A.J. Camm, Editors. 1995, Futura Publishing COmpany, Inc.: Armonk NY. p. 63-74.

150. Murphy, D.A., et al., *The heart reinnervates after transplantation.* Ann Thorac Surg, 2000. **69**(6): p. 1769-81.

151. Ramaekers, D., et al., *Heart rate variability after cardiac transplantation in humans.* Pacing Clin Electrophysiol, 1996. **19**(12 Pt 1): p. 2112-9.

152. Kember, G., et al., *Competition model for aperiodic stochastic resonance in a Fitzhugh-Nagumo model of cardiac sensory neurons.* Physical Review E, 2001. **63**(4 Pt 1): p. 041911.

153. Kember, G.C., et al., *Aperiodic stochastic resonance in a hysteretic population of cardiac neurons.* Physical Review E, 2000. **61**(2): p. 1816-1824.

154. Berntson, G.G., et al., *Heart rate variability: origins, methods, and interpretive caveats.* Psychophysiology, 1997. **34**(6): p. 623-48.

155. Huikuri, H.V., et al., *Circadian rhythms of frequency domain measures of heart rate variability in healthy subjects and patients with coronary artery disease. Effects of arousal and upright posture.* Circulation, 1994. **90**(1): p. 121-6.

156. Singh, R.B., et al., *Circadian heart rate and blood pressure variability considered for research and patient care.* Int J Cardiol, 2003. **87**(1): p. 9-28; discussion 29-30.

157. Bernardi, L., et al., *Physical activity influences heart rate variability and very-low-frequency components in Holter electrocardiograms.* Cardiovasc Res, 1996. **32**(2): p. 234-7.

158. Stein, P.K., et al., *Traditional and nonlinear heart rate variability are each independently associated with mortality after myocardial infarction.* J Cardiovasc Electrophysiol, 2005. **16**(1): p. 13-20.

159. Kleiger, R.E., et al., *Decreased heart rate variability and its association with increased mortality after acute myocardial infarction.* American Journal of Cardiology, 1987. **59**(4): p. 256-262.

160. Damasio, A., *Looking for Spinoza: Joy, Sorrow, and the Feeling Brain* 2003, Orlando: Harcourt.

161. Stein, J., ed. *The Random House College Dictionary.* 1975, Random House: New York. 261.

162. Strogatz, S. and I. Stewart, *Coupled Oscillators and Biological Synchronization.* Scientific American, 1993(December): p. 102-109.

163. Tiller, W.A., R. McCraty, and M. Atkinson, *Cardiac coherence: A new, noninvasive measure of autonomic nervous system order.* Alternative Therapies in Health and Medicine, 1996. **2**(1): p. 52-65.

164. Bradley, R.T. and K.H. Pribram, *Communication and stability in social collectives.* Journal of Social and Evolutionary Systems, 1998. **21**(1): p. 29-80.

165. Ho, M.-W., *The Rainbow and the Worm: The Physics of Organisms* 2005, Singapore: World Scientific Publishing Co.

166. Leon, E., et al., *Affect-aware behavior modelling and control inside an intelligent environment* Pervasive and Mobile Computing doi:10.1016/j.pmcj.2009.12.002, 2010. 167. Hasan, Y., L. Begue, and B.J. Bushman, *Violent video games stress people out and make them more aggressive.* Aggress Behav, 2013. **39**(1): p. 64-70.

168. McCraty, R., et al., *The effects of emotions on short-term power spectrum analysis of heart rate variability.* American Journal of Cardiology, 1995. **76**(14): p. 1089-1093.

169. Pribram, K.H. and F.T. Melges, *Psychophysiological basis of emotion*, in *Handbook of Clinical Neurology*, P.J. Vinken and G.W. Bruyn, Editors. 1969, North-Holland Publishing Company: Amsterdam. p. 316-341.

170. Bradley, R.T., *Charisma and Social Structure: A Study of Love and Power, Wholeness and Transformation* 1987, New York: Paragon House.

171. LeDoux, J., *The Emotional Brain: The Mysterious Underpinnings of Emotional Life* 1996, New York: Simon and Schuster.

172. Miller, G.A., E.H. Galanter, and K.H. Pribram, *Plans and the Structure of Behavior* 1960, New York: Henry Holt & Co.

173. Pribram, K.H., *Feelings as monitors*, in *Feelings and Emotions*, M.B. Arnold, Editor 1970, Academic Press: New York. p. 41-53.

174. Olatunji, B.O., et al., *Heightened attentional capture by threat in veterans with PTSD.* J Abnorm Psychol, 2013. **122**(2): p. 397-405.

175. Pribram, K.H. and D. McGuinness, *Arousal, activation, and effort in the control of attention.* Psychological Review, 1975. **82**(2): p. 116-149.

176. Pribram, K.H., *Languages of the Brain: Experimental Paradoxes and Principals in Neuropsychology* 1971, New York: Brandon House.

177. Frysinger, R.C. and R.M. Harper, *Cardiac and respiratory correlations with unit discharge in epileptic human temporal lobe.* Epilepsia, 1990. **31**(2): p. 162-171.

178. Childre, D.L., *Freeze-Frame®, Fast Action Stress Relief* 1994, Boulder Creek: Planetary Publications.

179. Childre, D. and H. Martin, *The HeartMath Solution* 1999, San Francisco: HarperSanFrancisco.

180. Childre, D. and B. Cryer, *From Chaos to Coherence: The Power to Change Performance* 2000, Boulder Creek, CA: Planetary.

181. Childre, D. and D. Rozman, *Overcoming Emotional Chaos: Eliminate Anxiety, Lift Depression and Create Security in Your Life* 2002, San Diego: Jodere Group.

182. Childre, D. and D. Rozman, *Transforming Stress: The HeartMath Solution to Relieving Worry, Fatigue, and Tension* 2005, Oakland, CA: New Harbinger Publications.

183. Lehrer, P., et al., *Heart rate variability biofeedback: effects of age on heart rate variability, baroreflex gain, and asthma.* Chest, 2006. **129**(2): p. 278-84.

184. Ratanasiripong, P., N. Ratanasiripong, and D. Kathalae, *Biofeedback Intervention for Stress and Anxiety among Nursing Students: A Randomized Controlled Trial.* International Scholarly Research Network Nurs. 2012;2012:827972, 2012.

185. Beckham, A.J., T.B. Greene, and S. Meltzer-Brody, *A pilot study of heart rate variability biofeedback therapy in the treatment of perinatal depression on a specialized perinatal psychiatry inpatient unit.* Arch Womens Ment Health, 2013. **16**(1): p. 59-65.

186. Siepmann, M., et al., *A pilot study on the effects of heart rate variability biofeedback in patients with depression and in healthy subjects.* Appl Psychophysiol Biofeedback, 2008. **33**(4): p. 195-201.

187. Hallman, D.M., et al., *Effects of heart rate variability biofeedback in subjects with stress-related chronic neck pain: a pilot study.* Appl Psychophysiol Biofeedback, 2011. **36**(2): p. 71-80.

188. Henriques, G., et al., *Exploring the effectiveness of a computer-based heart rate variability biofeedback program in reducing anxiety in college students.* Appl Psychophysiol Biofeedback, 2011. **36**(2): p. 101-12.

189. Lin, G., et al., *Heart rate variability biofeedback decreases blood pressure in prehypertensive subjects by improving autonomic function and baroreflex.* J Altern Complement Med, 2012. **18**(2): p. 143-52.

190. McCraty, R. and D. Tomasino, *Coherence-building techniques and heart rhythm coherence feedback: New tools for stress reduction, disease prevention, and rehabilitation*, in *Clinical Psychology and Heart Disease*, E. Molinari, A. Compare, and G. Parati, Editors. 2006, Springer-Verlag: Milan, Italy.

191. Li, W.-C., et al., *The Investigation of Visual Attention and Workload by Experts and Novices in the Cockpit*, in *Engineering Psychology and Cognitive Ergonomics. Applications and Services*, D. Harris, Editor 2013, Springer Berlin Heidelberg. p. 167-176.

192. Luskin, F., K. Newell, and W. Haskell, *Stress management training of elderly patients with congestive heart failure: pilot study.* Preventive Cardiology, 1999. **2**: p. 101-104.

193. Weltman, G., et al., *Police Department Personnel Stress Resilience Training: An Institutional Case Study.* Global Advances in Health and Medicne, 2014. **3**(2): p. 72-79.

194. Lehrer, P., Y. Sasaki, and Y. Saito, *Zazen and cardiac variability.* Psychosomatic Medicine, 1999. **61**: p. 812-821.

195. Kim, D.-K., et al., *Dynamic correlations between heart and brain rhythm during Autogenic meditation.* Front. Hum. Neurosci., 2013. **7:414**.

196. Peng, C.K., et al., *Exaggerated heart rate oscillations during two meditation techniques.* Int J Cardiol, 1999. **70**(2): p. 101-7.

197. Wu, S.D. and P.C. Lo, *Inward-attention meditation increases parasympathetic activity: a study based on heart rate variability.* Biomed Res, 2008. **29**(5): p. 245-50.

198. Phongsuphap, S. and Y. Pongsupap, *Analysis of Heart Rate Variability during Meditation by a Pattern Recognition Method* Computing in Cardiology, 2011. **38**: p. 197-200.

199. Stanley, R., *Types or prayer, heart rate variablity and innate healing* Zygon, 2009. **44**(4): p. 825-846.

200. Bernardi, L., et al., *Effect of rosary prayer and yoga mantras on autonomic cardiovascular rhythms: Comparative study.* BMJ, 2001. **323**: p. 1446-1449.

201. Stanley, R., *Types of Prayer, Heart Rate Variablity and Innate Healing* Zygon 2009. **44**(4).

202. Lehrer, P., et al., *Effects of rhythmical muscle tension at 0.1Hz on cardiovascular resonance and the baroreflex.* Biol Psychol, 2009. **81**(1): p. 24-30.

203. Baule, G. and R. McFee, *Detection of the magnetic field of the heart.* American Heart Journal, 1963. **55**(7): p. 95-96.

204. Nakaya, Y., *Magnetocardiography: a comparison with electrocardiography.* J Cardiogr Suppl, 1984. **3**: p. 31-40.

205. Halberg, F., et al., *Cross-spectrally coherent ~10.5- and 21-year biological and physical cycles, magnetic storms and myocardial infarctions.* Neuroendocrinology, 2000. **21**: p. 233-258.

206. Pribram, K.H., *Brain and Perception: Holonomy and Structure in Figural Processing* 1991, Hillsdale, NJ: Lawrence Erlbaum Associates, Publishers.

207. Prank, K., et al., *Coding of time-varying hormonal signals in intracellular calcium spike trains.* Pac Symp Biocomput, 1998: p. 633-44.

208. Schofl, C., K. Prank, and G. Brabant, *Pulsatile hormone secretion for control of target organs.* Wiener Medizinische Wochenschrift, 1995. **145**(17-18): p. 431-435.

209. Schofl, C., et al., *Frequency and amplitude enhancement of calcium transients by cyclic AMP in hepatocytes.* Biochem J, 1991. **273**(Pt 3): p. 799-802.

210. Coles, M.G.H., G. Gratton, and M. Fabini, *Event-related brain potentials*, in *Principles of Psychophysiology: Physical, Social and Inferential Elements*, J.T. Cacioppo and L.G. Tassinary, Editors. 1990, Cambridge University Press: NY.

211. Song, L.Z., G.E. Schwartz, and L.G. Russek, *Heart-focused attention and heart-brain synchronization: Energetic and physiological mechanisms.* Alternative Therapies in Health and Medicine, 1998. **4**(5): p. 44-62.

212. McCraty, R., M. Atkinson, and W.A. Tiller, *New electrophysiological correlates associated with intentional heart focus.* Subtle Energies, 1993. **4**(3): p. 251-268.

213. Russell, P., *The Brain Book* 1979, New York: Penguin Books USA.

214. Hatfield, E., *Emotional Contagion* 1994, New York: Cambridge University Press.

215. McCraty, R., et al. *The Electricity of Touch: Detection and measurment of cardiac energy exchange between people.* in *The Fifth Appalachian Conference on Neurobehavioral Dynamics: Brain and Values*. 1996. Radford VA: Lawrence Erlbaum Associates, Inc. Mahwah, NJ.

216. Anshel, M.H., *Effect of chronic aerobic exercise and progressive relaxation on motor performance and affect following acute stress.* Behav Med, 1996. **21**(4): p. 186-96.

217. Stroink, G., *Principles of cardiomagnetism*, in *Advances in Biomagnetism*, S.J. Williamson, et al., Editors. 1989, Plenum Press: New York. p. 47-57.

218. Anshel, M., *A conceptual model and implications for coping wtih stressful events in police work.* Criminal Justice and Behavior, 2000. **27**(3): p. 375-400.

219. McCraty, R., *Influence of cardiac afferent input on heart-brain synchronization and cognitive performance.* International Journal of Psychophysiology, 2002. **45**(1-2): p. 72-73.

220. Holcomb, B.K., et al., *Dimethylamino parthenolide enhances the inhibitory effects of gemcitabine in human pancreatic cancer cells.* J Gastrointest Surg, 2012. **16**(7): p. 1333-40.

221. Morris, S.M., *Facilitating collective coherence: Group Effects on Heart Rate Variability Coherence and Heart Rhythm Synchronization.* Alternative Therapies in Health and Medicine, 2010. **16**(4): p. 62-72.

222. Waters, J.A., et al., *Single-port laparoscopic right hemicolectomy: the first 100 resections.* Dis Colon Rectum, 2012. **55**(2): p. 134-9.

223. Nelson, R. *Scientific Evidence for the Existence of a True Noosphere: Foundation for a Noo-Constitution.* in *World Forum of Spiritual Culture.* 2010. Astana, Kazakhstan.

224. Hodgkinson, G.P., J. Langan-Fox, and E. Sadler-Smith, *Intuition: A fundamental bridging construct in the behavioural sciences.* British Journal of Psychology, 2008. **99**(1): p. 1-27.

225. Myers, D.G., *Intuition: Its Powers and Perils* 2002, New Haven: Yale University Press.

226. Bradley, R.T., et al., *Nonlocal Intuition in Entrepreneurs and Non-entrepreneurs: Results of Two Experiments Using Electrophysiological Measures.* International Journal of Entrepreneurship and Small Business, 2011. **12**(3): p. 343-372.

227. Dane, E. and M.G. Pratt, *Exploring intuition and its role in managerial decision making.* Academy of Management Review, 2007. **32**: p. 33–54.

228. Bastick, T., *Intuition: How we think and act* 1982, New York:: Wiley.

229. Moir, A. and D. Jessel, *Brainsex: The real difference between men and women* 1989, London:: Mandarin Paperbacks.

230. Larsen, A. and C. Bundesen, *A template-matching pandemonium recognizes unconstrained handwritten characters with high accuracy.* Mem Cognit, 1996. **24**(2): p. 136-43.

231. Craig, J. and N. Lindsay, *Quantifying "gut feeling" in the opportunity recognition process.* Frontiers of Entrepreneurship Research, 2001: p. 124-135.

232. Halberg, F., et al., *Time Structures (Chronomes) of the Blood Circulation, Populations' Health, Human Affairs and Space Weather.* World Heart Journal, 2011. **3**(1): p. 1-40.

233. Uyeda, S., et al., *Geoelectric potential changes: possible precursors to earthquakes in Japan.* Proc Natl Acad Sci U S A, 2000. **97**(9): p. 4561-6.

234. Wiseman, R. and M. Schlitz, *Experimenter effects and the remote detection of staring.* Journal of Parapsychology, 1997. **61**: p. 197-207.

235. Bohm, D. and B.J. Hiley, *The Undivided Universe* 1993, London: Routledge.

236. Laszlo, E., *The Interconnected Universe: Conceptual Foundations of Transdiciplinary Unified Theroy* 1995, Singapore: World Scientific.

237. Nadeau, R. and M. Kafatos, *The Non-Local Universe: The New Physics and Matters of the Mind* 1999, New York: Oxford University Press.

238. Mayer, R.E., *The search for insight: Grappling with gestalt psychology's unanswered questions.*, in *The nature of insight*, R.J. Sternberg and J.E. Davidson, Editors. 1996, The MIT Press: Cambridge, MA. p. 3–32.

239. Hogarth, R.M., *Educating Intuition* 2001, Chicago: The University of Chicago Press.

240. Bem, D.J., *Feeling the future: Experimental evidence for anomalous retroactive influences on cognition and affect.* J Pers Soc Psychol, 2011.

241. Radin, D., *The Conscious Universe: The Scientific Truth of Psychic Phenomena* 1997, San Francisco, CA: HarperEdge.

242. Mossbridge, J., P. Tressoldi, E, and J. Utts *Predictive Physiological Anticipation Preceding Seemingly Unpredictable Stimuli: A Meta-Analysis.* Frontiers in Psychology, 2012. **3**:390.

243. McCraty, R., M. Atkinson, and R.T. Bradley, *Electrophysiological evidence of intuition: Part 1. The surprising role of the heart.* Journal of Alternative and Complementary Medicine, 2004. **10**(1): p. 133-143.

244. McCraty, R., M. Atkinson, and R.T. Bradley, *Electrophysiological evidence of intuition: Part 2. A system-wide process?* Journal of Alternative and Complementary Medicine, 2004. **10**(2): p. 325-336.

245. Tressoldi, P.E., et al., *Heart rate differences between targets and non targets in intuition tasks.* Fiziol Cheloveka, 2005. **31**(6): p. 32-6.

246. Hu, H. and M. Wu, *New Nonlocal Biological Effect.* NeuroQuantology 2012. **10**(3): p. 462-467.

247. Tressoldi, P.E., et al., *Implicit Intuition: How Heart Rate can Contribute to Prediction of Future Events.* Journal of the Society for Psychical research 2009. **73**: p. 1-16.

248. Sartori, L., et al., *Physiological correlates of ESP: heart rate differences between targets and nontargets.* Journal of Parapsychology, 2004. **68**(2): p. 351.

249. Tressoldi, P.E., et al., *Further evidence of the possibility of exploiting anticipatory physiological signals to assist implicit intuition of random events.* Journal of Scientific Exploration, 2010. **24**(3): p. 411.

250. Bradley, R.T., R. McCraty, M. Atkinson, & M. Gillin. *Nonlocal Intuition in Entrepreneurs and Nonentrepreneurs: An Experimental Comparison Using Electrophysiological Measures.* in *Regional Frontiers of Entrepreneurship Research.* 2008. Hawthorne, Australia.

251. Toroghi, S.R., et al., *Nonlocal Intuition: Replication and Paired-Subjects Enhancement Effects.* Global Advances in Health and Medicne, 2014.

252. McCraty, R., *Electrophysiology of Intuition: Pre-stimulus Responses in Group and Individual Participants Using a Roulette*

Paradigm. Global Advances in Health and Medicne, 2014. **3**(2): p. 16-27.

253. Laszlo, E., *Quantum Shift in the Global Brain: how the new scientific reality can change us and our world* 2008, Rochester, VT: Inner Traditions.

254. Mitchell, E., *Quantum holography: a basis for the interface between mind and matter*, in *Bioelectromagnetic Medicine*, P.G. Rosch and M.S. Markov, Editors. 2004, Dekker: New York, NY. p. 153-158.

255. Tiller, W.A., J. W E Dibble, and M.J. Kohane, *Conscious Acts of Creation: The Emergence of a New Physics* 2001, Walnut Creek, CA: Pavior Publishing. (pp. 201-202).

256. Bradley, R.T., *Psycholphysiology of Intution: A quantum-holgraphic theory on nonlocal communication.* World Futures: The Journal of General Evolution, 2007. **63**(2): p. 61-97.

257. Marcer, P. and W. Schempp, *The brain as a conscious system.* Internationl Journal of General Systems, 1998. **27**: p. 231-248.

258. Pribram, K.H. and R.T. Bradley, *The brain, the me and the I*, in *Self-Awareness: Its Nature and Development*, M. Ferrari and R. Sternberg, Editors. 1998, The Guilford Press: New York. p. 273-307.

259. Schempp, W., *Quantum holograhy and neurocomputer architectures.* Journal of Mathematical Imaging and vision, 1992. **2**: p. 109-164.

260. Simons, D.J. and C.F. Chabris, *Gorillas in our midst: Sustained inattentional blindness for dynamic events.* Perception, 1999. **28**(9): p. 1059-1074.

261. Baumeister, R.F., *Ego depletion and self-regulation failure: a resource model of self-control.* Alcohol Clin Exp Res, 2003. **27**(2): p. 281-4.

262. Petitmengin-Peugeot, C., *The Intuitive Experience*, in *The View from Within. First-person approaches to the study of consciousness*, F.J.Varela and J. Shear, Editors. 199, Imprint Academic: London,. p. 43-77.

263. Cutler, J.A., et al., *An overview of randomized trials of sodium reduction and blood pressure.* Hypertension, 1991. **17**(1 Suppl): p. I27-133.

264. MacMahon, S., et al., *Obesity and hypertension: epidemiological and clinical issues.* Eur Heart J, 1987. **8 Suppl B**: p. 57-70.

265. MacMahon, S., et al., *Blood pressure, stroke, and coronary heart disease. Part 1, Prolonged differences in blood pressure: prospective observational studies corrected for the regression dilution bias.* Lancet, 1990. **335**(8692): p. 765-774.

266. McCraty, R., et al., *New Hope for Correctional Officers: An Innovative Program for Reducing Stress and Health Risks.* Appl Psych and Biofeedback 2009. **34**(4): p. 251-272.

267. Lehrer, P., et al., *Biofeedback treatment for asthma.* Chest, 2004. **126**(2): p. 352-361.

268. Lehrer, P., Carr, RE., Smetankine, A., Vaschillo, E., Peper, E., Porges, S., Edelberg, R., Hamer, R., Hochron, S., *Respiratory sinus arrhythmia versus neck/trapezius EMG and incentive inspirometry biofeedback for asthma: a pilot study.* Applied Psychophysiology & Biofeedback, 1997. **22**(2): p. 95-109.

269. Lehrer, P.M., E. Vaschillo, and B. Vaschillo, *Resonant frequency biofeedback training to increase cardiac variability. Rationale and manual for training.* Applied Psychophyisology and Biofeedack, 2000. **25**(3): p. 177-191.

270. Karavidas, M., *Psychophysiological Treatment for Patients with Medically Unexplained Symptoms: A Randomized Controlled Trial.* Psychosomatics, in press.

271. Hassett, A.L., et al., *A pilot study of the efficacy of heart rate variability (HRV) biofeedback in patients with fibromyalgia.* Appl Psychophysiol Biofeedback, 2007. **32**(1): p. 1-10.

272. Karavidas, M.K., et al., *Preliminary results of an open label study of heart rate variability biofeedback for the treatment of major depression.* Appl Psychophysiol Biofeedback, 2007. **32**(1): p. 19-30.

273. McCraty, R., M. Atkinson, and L. Lipsenthal, *Emotional self-regulation program enhances psychological health and quality of life in patients with diabetes.* Boulder Creek, CA: HeartMath Research Center, HeartMath Institute, Publication No. 00-006., 2000.

274. Bradley, R.T., McCraty, R., Atkinson, M., Tomasino., D., *Emotion Self-Regulation, Psychophysiological Coherence, and Test Anxiety: Results from an Experiment Using Electrophysiological Measures.* Applied Psychophysiology and Biofeedback, 2010. **35**(4): p. 261-283.

275. Luskin, F., et al., *A controlled pilot study of stress management training of elderly patients with congestive heart failure.* Preventive Cardiology, 2002. **5**(4): p. 168-172, 176.

276. Arguelles, L., R. McCraty, and R.A. Rees, *The heart in holistic education.* Encounter: Education for Meaning and Social Justice, 2003. **16**(3): p. 13-21.

277. Barrios-Choplin, B., R. McCraty, and B. Cryer, *An inner quality approach to reducing stress and improving physical and emotional wellbeing at work.* Stress Medicine, 1997. **13**(3): p. 193-201.

278. McCraty, R., *Heart-brain neurodynamics: The making of emotions* 2003, Boulder Creek, CA: HeartMath Research Center, HeartMath Institute, Publication No. 03-015.

279. McCraty, R. and M. Atkinson, *Spontaneous heart rhythm coherence in individuals practiced in positive-emotion-focused techniques.* Unpublished data, 1998.

280. McCraty, R., et al., *Impact of the Power to Change Performance program on stress and health risks in correctional officers* 2003: Boulder Creek, CA: HeartMath Research Center, HeartMath Institute, Report No. 03-014, November 2003.

281. Nada, P.J., *Heart rate variability in the assessment and biofeedback training of common mental health problems in children.* Med Arh, 2009. **63**(5): p. 244-8.

282. Bradford, E.J., K.A. Wesnes, and D. Brett, *Effects of peak performance training on cognitive function.* Journal of Psychopharmacology, 2005. **19**(5 suppl): p. A44.

283. Kim, S., et al., *Heart rate variability biofeedback, executive functioning and chronic brain injury.* Brain Inj, 2013. **27**(2): p. 209-22.

284. Berry, M.E., et al., *Non-pharmacological Intervention for Chronic Pain in Veterans: A Pilot Study of Heart Rate Variability Biofeedback.* Global Advances in Health and Medicne, 2014. **3**(2): p. 28-33.

285. Soer, R., et al., *Heart Coherence Training Combined with Back School in Patients with Chronic Non-specific Low Back Pain: First Pragmatic Clinical Results.* Appl Psychophysiol Biofeedback, 2014.

286. Scott, L.D., W.-T. Hwang, and A.E. Rogers, *The impact of multiple care giving roles on fatigue, stress, and work performance among hospital staff nurses.* Journal of Nursing Administration, 2006. **36**(2): p. 86-95.

287. Salmond, S. and P.E. Ropis, *Job stress and general well-being: a comparative study of medical-surgical and home care nurses.* Medsurg Nursing, 2005. **14**(5): p. 301.

288. Pipe, T. and J. Bortz, *Mindful leadership as healing practice: Nurturing self to serve others.* International Journal for Human Caring, 2009. **13**(2): p. 34-38.

289. Sarabia-Cobo, C., *Heart Coherence: A New Tool in the Management of Stress on Professionals and Family Caregivers of Patients with Dementia.* Applied Psychophysiology and Biofeedback, 2015: p. 1-9.

290. Watson, J., *Nursing: Human science and human care: A theory of nursing* 1999: Jones & Bartlett Learning.

291. Lemaire, J.B., Wallllace J E, Lewin A M, de Grood J, Schaefer J P, *The effect of a biofeedback-based stress management tool on physician stress: a randomized controlled clinical trial.* Open Medicine, 2011. **5**(4): p. 154-163.

292. HeartMath, L.L.C., *Return on Investment.* White Paper, 2009.

293. Reissner, A., *The dance of partnership: A theological reflection.* Missiology: An International Review, 2001. **29**(1): p. 3-10.

294. Nahser, F. and S. Mehrtens, *What's Really Going On?* 1993, Chicago: Corporantes.

295. Goldman, L., *Breaking the Silence: A Guide to Helping Children with Complicated Grief-Suicide, Homicide, AIDS, Violence and Abuse* 2014: Routledge.

296. Perry, B.D., *Childhood experience and the expression of genetic potential: What childhood neglect tells us about nature and nurture.* Brain and mind, 2002. **3**(1): p. 79-100.

297. Costello, E.J., et al., *Psychiatric disorders in pediatric primary care. Prevalence and risk factors [see comments].* Arch Gen Psychiatry, 1988. **45**(12): p. 1107-16.

298. Scales, P.C., *Reducing risks and building developmental assets: Essential actions for promoting adolescent health.* Journal of School Health, 1999. **69**(3): p. 113-119.

299. Bennett, W., *The Index of Leading Cultural Indicators: Facts and Figures on the State of American Society* 1994, New York: Simon & Schuster.

300. Resnick, M.D., L.J. HARRIS, and R.W. Blum, *The impact of caring and connectedness on adolescent health and well-being.* J Paediatr Child Health, 1993. **29**(s1): p. S3-S9.

301. Bradley, R.T., et al., *Reducing Test Anxiety and Improving Test Performance in America's Schools: Results from the TestEdge National Demonstration Study* 2007, Boulder Creek, CA: HeartMath Research Center, HeartMath Institute, Publication No. 07-09-01.

302. Hartnett-Edwards, K. and T.C.G. University, *The Social Psychology and Physiology of Reading/language Arts Achievement* 2006: Claremont Graduate University.

303. Connolly, F., *Evaulation of a HeartMath / Safe Place Programme with Sshool Childern in West Belfast*, 2009, Greater Falls Extended Schools: http://taketen.tv/file/fccBrochure.pdf. p. 1-12.

304. Bradley, R.T., et al., *Efficacy of an Emotion Self-regulation Program for Promoting Development in Preschool Children.* Glob Adv Health Med, 2012. **1**(1): p. 36-50.

305. May, R.W., M.A. Sanchez-Gonzalez, and F.D. Fincham, *School burnout: increased sympathetic vasomotor tone and attenuated ambulatory diurnal blood pressure variability in young adult women.* Stress, 2014(0): p. 1-9.

306. May, R.W., K.N. Bauer, and F.D. Fincham, *School Burnout: Diminished Academic and Cognitive Performance. Learning and Individual Differences.* . In review

307. Bajkó, Z., et al., *Anxiety, depression and autonomic nervous system dysfunction in hypertension.* J Neurol Sci, 2012. **317**(1): p. 112-116.

308. FitzGerald, L., et al., *Effects of dipping and psychological traits on morning surge in blood pressure in healthy people.* Journal of Human Hypertension, 2012. **26**(4): p. 228-235.

309. Matthews, K.A., et al., *Blood pressure reactivity to psychological stress predicts hypertension in the CARDIA study.* Circulation, 2004. **110**(1): p. 74-78.

310. Unsworth, N., et al., *An automated version of the operation span task.* Behavior research methods, 2005. **37**(3): p. 498-505.

311. Unsworth, N., et al., *Complex working memory span tasks and higher-order cognition: A latent-variable analysis of the relationship between processing and storage.* Memory, 2009. **17**(6): p. 635-654.

312. Babraj, J.A., et al., *Extremely short duration high intensity interval training substantially improves insulin action in young healthy males.* BMC Endocrine Disorders, 2009. **9**(1): p. 3.

313. Rakobowchuk, M., et al., *Sprint interval and traditional endurance training induce similar improvements in peripheral arterial stiffness and flow-mediated dilation in healthy humans.* American Journal of Physiology-Regulatory, Integrative and Comparative Physiology, 2008. **295**(1): p. R236-R242.

314. Patchell, B., *Coherent Learning: Creating High-level Performance and Cultural Empathy From Student to Expert.* Global Advances in Health and Medicine, 2014. **3**(Suppl 1): p. BPA17.

315. Vislocky, M. and R. Leslie, *Efficacy and Implementation of HeartMath Instruction in College Readiness Program: Improving Students' Mathematics Performance and Learning* 2005, University of Cincinnati – Clermont College, Batavia OH: http://mathematics.clc.uc.edu/Vislocky/CPR%20Project.htm.

316. deBoer, R.W., J.M. Karemaker, and J. Strackee, *Hemodynamic fluctuations and baroreflex sensitivity in humans: A beat-to-beat model.* American Journal of Physiology, 1987. **253**(3 Pt 2): p. H680-H689.

317. Association, A.P., *Stress in America: findings* 2010.

318. Hoel, H., K. Sparks, and C.L. Cooper, *The cost of violence/stress at work and the benefits of a violence/stress-free working environment.* Geneva: International Labour Organization, 2001.

319. Kalia, M., *Assessing the economic impact of stress [mdash] The modern day hidden epidemic.* Metabolism, 2002. **51**(6): p. 49-53.

320. Bliss, W.G., *Cost of employee turnover.* The Advisor, 2004.

321. Cooper, C. and R. Payne, eds. *Causes, Coping and Consequences of Stress at Work.* 1988, John Wiley & Sons Ltd.: New York.

322. Goetzel, R.Z., et al., *The relationship between modifiable health risks and health care expenditures. An analysis of the multi-employer HERO health risk and cost database. The Health Enhancement Research Organization (HERO) Research Committee.* Journal of Occupational and Environmental Medicine, 1998. **40**(10): p. 843-854.

323. Bosma, H., et al., *Low job control and risk of coronary heart disease in Whitehall II (prospective cohort) study.* Bmj, 1997. **314**(7080): p. 558-65.

324. Berkman, L.F. and S.L. Syme, *Social networks, host resistance, and mortality: a nine-year follow-up study of Alameda County residents.* Am J Epidemiol, 1979. **109**(2): p. 186-204.

325. Hermes, G.L., et al., *Social isolation dysregulates endocrine and behavioral stress while increasing malignant burden of spontaneous mammary tumors.* Proc Natl Acad Sci U S A, 2009. **106**(52): p. 22393-8.

326. Marmot, M.G. and S.L. Syme, *Acculturation and coronary heart disease in Japanese-Americans.* Am J Epidemiol, 1976. **104**(3): p. 225-47.

327. Neser, W., H. Tyroler, and J. Cassel, *Social disorganization and stroke mortality in the black population of North Carolina.* American Journal of Epidemiology, 1971. **93**(3): p. 166-175.

328. Ornstein, R. and D. Sobel, *The Healing Brain* 1987, New York: Simon and Schuster.

329. Lynch, J.J., *A Cry Unheard: New Insights into the Medical Consequences of Loneliness* 2000, Baltimore, MD: Bancroft Press.

330. Uchino, B.N., J.T. Cacioppo, and J.K. Kiecolt-Glaser, *The relationship between social support and physiological processes: a review with emphasis on underlying mechanisms and implications for health.* Psychol Bull, 1996. **119**(3): p. 488-531.

331. Cohen, S. and S. Syme, eds. *Social Support and Health.* 1985, Academic Press: Orlando.

332. Ornish, D., *Love and Survival: The Scientific Basis for the Healing Power of Intimacy* 1998, New York: HarperCollins Publishers.

333. Pipe, T.B., et al., *Building personal and professional resources of resilience and agility in the healthcare workplace.* Stress and Health, 2012. **28**(1): p. 11-22.

334. Newsome, M., et al., *Changing Job Satisfaction, Absenteeism, and Healthcare Claims Costs In a Hospital Culture.* Global Advances in Health and Medicine, 2014. **3**(Suppl 1): p. BPA01.

335. Riley, K. and D. Gibbs, *HeartMath in UK healthcare: Does it add up?* Journal of holistic healthcare, 2013. **10**(1): p. 23-28.

336. Murphy, H., *Caring Theory and HeartMath: A Match Made in Heaven.* Global Advances in Health and Medicine, 2014. **3**(Suppl 1): p. BPA18.

337. Goldfisher, A.M., B. Hounslow, and J. Blank, *Transforming and Sustaining the Care Environment.* Global Advances in Health and Medicine, 2014. **3**(Suppl 1): p. BPA11.

338. Bosteder, L. and S. Hargrave, *Learning within a Prison Environment: Will Emotional Intelligence Training Benefit Female Inmates Participating in a Work-based Education Program?*, 2008, Oregon State University: https://www.heartmath.org/research/research-library/educational/learning-within-a-prison-environment/. p. 1-3.

339. McCraty, R., A. Deyhle, and D. Childre, *The global coherence initiative: creating a coherent planetary standing wave.* Glob Adv Health Med, 2012. **1**(1): p. 64-77.

340. Uyeda, S., et al., *Geoelectric potential changes: possible precursors to earthquakes in Japan.* Proc Natl Acad Sci U S A, 2000. **97**(9): p. 4561-6.

341. Kopytenko, Yu A., et al. "Detection of ultra-low-frequency emissions connected with the Spitak earthquake and its aftershock activity, based on geomagnetic pulsations data at Dusheti and Vardzia observatories." Physics of the Earth and Planetary Interiors **77.1** (1993): p. 85-95

342. Cornelissen, G., et al., *Chronomes, Time Structures, for Chronobioengineering for "A Full Life".* Biomedical Instrumentation and Technology, 1999. **33**: p. 152-187.

343. Doronin, V.N., Parfentev, V.A., Tleulin, S.Zh, .Namvar, R.A., Somsikov, V.M., Drobzhev, V.I. and Chemeris, A.V., *Effect of variations of the geomagnetic field and solar activity on human physiological indicators.* Biofizika, 1998. **43**(4): p. 647-653.

344. Kay, R.W., *Geomagnetic Storms: Association with Incidence of Depression as Measured by Hospital Admission.* British Journal of Psychiatry, 1994. **164**: p. 403-409.

345. Mikulecký, M., *Solar activity, revolutions and cultural prime in the history of mankind.* Neuroendocrinology Letters, 2007. **28**(6): p. 749-756.

346. Burch, J.B., Reif, J.S., Yost, M.G. , *Geomagnetic disturbances are associated with reduced nocturnal excretion of a melatonin metabolite in humans.* Neuroscience Letters, 1999. **266**: p. 209-212.

347. Rapoport, S.I., Blodypakova, T.D., Malinovskaia, N.K., Oraevskii, V.N., Meshcheriakova, S.A., Breus, T.K. and Sosnovskii, A.M., , *Magnetic storms as a stress factor.* Biofizika, 1998. **43**(4): p. 632-639.

348. Pobachenko, S.V., Kolesnik, A. G., Borodin, A. S., Kalyuzhin, V. V., *The Contigency of Parameters of Human Encephalograms and Schumann Resonance Electromagnetic Fields Revealed in Monitoring Studies.* Complex Systems Biophysics, 2006. **51**(3): p. 480-483.

349. Persinger, M.A., *Sudden unexpected death in epileptics following sudden, intense, increases in geomagnetic activity: prevalence of effect and potential mechanisms.* Int J Biometeorol, 1995. **38**(4): p. 180-187.

350. Stoupel, E., *Sudden cardiac deaths and ventricular extrasystoles on days of four levels of geomagnetic activity.* J. Basic Physiol. Pharmacol., 1993. **4**(4): p. 357-366.

351. Belov, D.R., Kanunikov, I. E., and Kiselev, B. V., *Dependence of human EEG synchronization on the geomagnetic activity on the day of experiment.* Ross Fiziol. Zh Im I M Sechenova, 1998. **84**(8): p. 761–774.

352. Villoresi, G., Ptitsyna, N.G., Tiasto, M.I. and Iucci, N., *Myocardial infarct and geomagnetic disturbances: analysis of data on morbidity and mortality [In Russian].* Biofizika, 1998. **43**(4): p. 623-632.

353. Gordon, C., Berk, M. , *The effect of geomagnetic storms on suicide.* South African Psychiat Rev, 2003. **6**: p. 24-27.

354. Kay, R.W., *Schizophrenia and season of birth: relationship to geomagnetic storms.* Schiz Res, 2004. **66**: p. 7-20.

355. Malin, S.R.C.a.S., B.J., *Correlation between heart attacks and magnetic activity.* Nature, 1979. **277**: p. 646-648.

356. Nikolaev, Y.S., Rudakov, Y.Y., Mansurov, S.M. and Mansurova, L.G., *Interplanetary magnetic field sector structure and disturbances of the central nervous system activity.* Reprint N 17a, Acad. Sci USSR, IZMIRAN, Moscow, 1976: p. 29.

357. Oraevskii, V.N., Breus, T.K., Baevskii, R.M., Rapoport, S.I., Petrov, V.M., Barsukova, Zh.V., Gurfinkel' Iul, and Rogoza, A.T. , *Effect of geomagnetic activity on the functional status of the body.* Biofizika, 1998. **43**(5): p. 819-826.

358. Zaitseva, S.A.a.P., M. I., *Effect of solar and geomagnetic activity on population dynamics among residents of Russia [In Russian].* Biofizika, 1995. **40**(4): p. 861-864.

359. Persinger, M.A., *Wars and increased solar-geomagnetic activity: aggression or change in intraspecies dominance?* Percept Mot Skills, 1999. **88**(3 Pt 2): p. 1351-1355.

360. Kleimenova, N. and O. Kozyreva, *Daytime quasiperiodic geomagnetic pulsations during the recovery phase of the strong magnetic storm of May 15, 2005.* Geomagnetism and Aeronomy, 2007. **47**(5): p. 580-587.

361. Subrahmanyam, S., P. Narayan, and T. Srinivasan, *Effect of magnetic micropulsations on the biological systems – A bioenvironmental study.* International Journal of Biometeorology, 1985. **29**(3): p. 293-305.

362. Halberg, F., et al., *Cycles Tipping the Scale between Death and Survival (="Life").* Progress of Theoretical Physics Supplement 2008. **173**: p. 153-181.

363. Otsuka, K., et al., *Chronomics and "Glocal" (Combined Global and Local) Assessment of Human Life.* Progress of Theoretical Physics Supplement, 2008. **173**: p. 134-152.

364. Persinger, M.A., *Geopsychology and geopsychopathology: Mental processes and disorders associated with geochemical and geophysical factors.* Experientia, 1987. **43**: p. 92-104.

365. Dimitrova, S., Stoilova, I. and Cholakov, I., *Influence of Local Geomagnetic Storms on Arterial Blood Pressure.* Bioelectromagnetics, 2004. **25**: p. 408-414.

366. Hamer, J.R., *Biological entrainment of the human brain by low frequency radiation.* Northrop Space Labs, 1965: p. 65-199.

367. Rapoport, S.I., Malinovskaia, N.K., Oraevskii, V.N., Komarov, F.I., Nosovskii, A.M. and Vetterberg, L., , *Effects of disturbances of natural magnetic field of the Earth on melatonin production in patients with coronary heart disease.* Klin Med (Mosk), 1997. **75**(6): p. 24-26.

368. Ertel, S., *Space weather and revolutions: Chizhevsky's heliobiological claim scrutinized.* Studia Psychologica, 1996. **39**: p. 3-22.

369. Grigoryev, P., Rozanov, V., Vaiserman, A., Vladimirskiy, B., *Heliogeophysical factors as possible triggers of suicide terroristic acts.* Health, 2009. **1**(4): p. 294-297.

370. Smelyakov, S.V. *Tchijevsky's Disclosure: How the Solar Cycles Modulate the History*. http://www.ASTROTHEOS.COM 2006.

371. Tchijevsky, A.L., (de Smitt, V.P. translation), *Physical Factors of the Historical Process.* Cycles, 1971. **22**: p. 11-27.

372. Ertel, S., *Cosmophysical correlations of creative activity in cultural history.* Biophysics, 1998. **43**(4): p. 696-702.

373. Stoupel, E., et al., *Ambulatory blood pressure monitoring in patients with hypertension on days of high and low geomagnetic activity.* J Hum Hypertens, 1995. **9**(4): p. 293-4.

374. Anshel, M.H., *Effect of age, sex, and type of feedback on motor performance and locus of control.* Res Q, 1979. **50**(3): p. 305-17.

375. Cornelissen, G., et al. *Gender differences in circadian and extra-circadian aspects of heart rate variability (HRV).* in *1st International Workshop of The TsimTsoum Institute.* 2010. Krakow, Poland.

376. Oinuma, S., et al., *Graded response of heart rate variability, associated with an alteration of geomagnetic activity in a subarctic area.* Biomed Pharmacother, 2002. **56**(Suppl 2): p. 284s-288s..

377. Anshel, M.H. and D. Marisi, *Effect of music and rhythm on physical performance.* Res Q, 1978. **49**(2): p. 109-13.

378. McCraty, R., *The energetic heart: Bioelectromagnetic communication within and between people,* in *Bioelectromagnetic Medicine,* P.J. Rosch and M.S. Markov, Editors. 2004, Marcel Dekker: New York. p. 541-562.

379. Kemper, K.J. and H.A. Shaltout, *Non-verbal communication of compassion: measuring psychophysiological effects.* BMC Complement Altern Med, 2011. 11: p. 132..

380. Montagnier, L., et al., *Transduction of DNA information through water and electromagnetic waves.* arXiv preprint arXiv:1501.01620, 2014.

381. Persinger, M., *On the possible representation of the electromagnetic equivalents of all human memory within the earth's magnetic filed: Implications of theoretical biology.* Theoretical Biology Insights, 2008. **1**: p. 3-11.

382. Persinger, M.A., *On the possibility of directly accessing every human brain by electromagnetic induction of the fundamental alogorithms* Perceptual and Motor Skills, 1995. **80**: p. 791-799.

383. Davies, J.L., *Alleviating political violence through enhancing coherence in collective consciousness: Impact assessment analysis of the Lebanon war.* Dissertation Abstracts International, 1988. **49**(8): p. 2381A.

384. Hagelin, J., *The Power of the Collective.* Shift: At the Frontier of Consciousness, 2007. **15**: p. 16-20.

385. Hagelin, J.S., Orme-Johnson, D. W., Rainforth, M., Cavanaugh, K., & Alexander, C. N. , *Results of the National Demonstration Project to Reduce Violent Crime and Improve Governmental Effectiveness in Washington, D.C.* Social Indicators Research, 1999. **47**: p. 153-201.

386. Orme-Johnson, D.W., et al., *International Peace Project in the Middle East THE EFFECTS OF THE MAHARISHI TECHNOLOGY OF THE UNIFIED FIELD* The Journal of Conflict Resolution, 1988. **32**(4): p. 776-812.

387. Bancel, P., Nelson, R., *The GCP Event Experiment: Design, Analytical Methods, Results.* Journal of Scientific Exploration, 2008. **22**(3): p. 309-333.

388. Nelson, R., *Effects of Globally Shared Attention and Emotion.* Journal of Cosmology, 2011. **14**.

389. Wendt, H.W., *Mass emotions apparently affect nominally random quantum processes: interplanetary magnetic field polarity found critical, but how is causal path?,* 2002, Halberg Chronobiology Center, University of Minnesota: St. Paul.

390. Ameling, A., *Prayer: an ancient healing practice becomes new again.* Holist Nurs Pract, 2000. **14**(3): p. 40-8.

391. Gillum, F. and D.M. Griffith, *Prayer and spiritual practices for health reasons among American adults: the role of race and ethnicity.* J Relig Health. **49**(3): p. 283-95.

392. Schwartz, S.A. and L. Dossey, *Nonlocality, intention, and observer effects in healing studies: laying a foundation for the future.* Explore (NY). **6**(5): p. 295-307.